AEROSPACE
Facts & Figures
2002/2003

S0-AXK-265

Compiled by:
Economic Data Service
Aerospace Research Center
Aerospace Industries Association of America, Inc.

Director, Aerospace Research Center
David H. Napier

Designed by:
Kathy Keler Graphics

Published by:
Aerospace Industries Association of America, Inc.
1250 Eye Street, N.W. #1200
Washington, D.C. 20005-3924
202-371-8400
202-371-8470 Fax
AIA@AIA-aerospace.org

Acknowledgments

Air BP Lubricants

Air Transport Association of America

The Boyd Company

Catherine J. Napier, 1/19/22 - 11/22/02

Council of Economic Advisers

Export-Import Bank of the United States

Futron Corporation

General Aviation Manufacturers Association

Helicopter Association International

International Civil Aviation Organization

National Aeronautics and Space Administration

National Science Foundation

Office of Management and Budget

U.S. Department of Commerce (Bureau of Economic Analysis; Bureau of the Census; International Trade Administration)

U.S. Department of Defense (Air Force; Army; Comptroller; Directorate for Information Operations and Reports; Missile Defense Agency; Navy)

U.S. Department of Labor (Bureau of Labor Statistics)

U.S. Department of Transportation (Federal Aviation Administration, Office of Airline Information)

CONTENTS

FOREWORD

Perhaps more than any other sector of the economy, the aviation and aerospace industries were the most seriously hurt by the events of September 11. The impact on aviation was immediate. U.S. commercial air travel stopped. For the second time in aviation history, air traffic declined from year to year. Despite layoffs, reduced flights, grounded equipment, and deferred deliveries, airlines posted record losses. The impact on the aerospace sector was less immediate and perhaps more prolonged. Aerospace industry sales actually increased in 2001. But recognizing the reduced demand from civil customers, manufacturers announced layoffs exceeding 60,000. In the 12 months since, aerospace employment has fallen 93,000 or 12% to the lowest level since before 1953!

Fortunately, thanks to the foresight of the Congress and President Bush, a commission was chartered prior to 9/11 to study the aerospace industry and make recommendations for ways to ensure that the U.S. would remain a global leader for the 21st century. I had the honor of serving as a member of the commission, along with 11 other representatives from various sectors of the aerospace community. Together, we heard testimony from at least 60 leaders from industry, government and labor, traveled to Europe and Asia to meet our counterparts, and worked with the commission staff to prepare three interim reports throughout the year and a final report that was released in November 2002. The recommendations made by the Commission were wide-ranging and covered subjects as diverse as workforce issues, air traffic modernization, launch range improvements, and aeronautics research and development. The Commission's work could not have been initiated at a more fortuitous time, as aviation and aerospace industries require innovative policy change more than at any time in the last 50 years.

There are signs of a recovery in late 2003 or 2004. Passenger traffic is gradually returning to pre-9/11 levels. Congress quickly passed the Airline Stabilization Act to help commercial air carriers, gave the president increased authority to make trade treaties, and substantially increased defense research and procurement, and increased NASA's funding. In the first nine months of 2002, military orders increased six percent and helped partially offset lower commercial orders.

Historically, the aviation and aerospace industries have faced far worse challenges. After World War I, the embryonic aviation industry faced near extinction, as aircraft orders dropped 90 percent. Without a commercial base, aviation could not have developed further. National policy helped right the aviation industry, as the government determined it would send the mail on those tiny flying machines. With that initial boost, the fledgling industry began working with government to develop federal regulations to license pilots, build landing fields, certify aircraft and establish traffic control. Thanks to government foresight, the aviation industry established a safety record and convinced a nervous public that air travel could be undertaken by the average man. Once the air transportation industry had a commercial base, there was no stopping the rapid development of modern aircraft. The lessons the manufacturing industry learned during those years had an enormous impact on the events of World War II, as military aircraft dominated the field of battle again and again. And to think that all this turned on the decision by the government 25 years earlier to send the mail via air.

The challenges facing the aviation and aerospace industries today are serious, but not insurmountable. Industry, working closely with the government, can help forge new policies that will propel the next generation of aircraft to maintain our economic health and very importantly, to protect our national security through the 21st century.

John W. Douglass
President and
Chief Executive Officer,
Aerospace Industries Association

AEROSPACE SUMMARY

U.S. aerospace industry's sales increased in 2001 to near record levels. However, negative indicators include: declining orders, profits, R&D and capital equipment investment; airlines posting record losses with several in or on the brink of bankruptcy; record-high imports; and record-low total employment, employment of R&D-performing scientists and engineers, and military aircraft production and exports. Obviously, serious challenges lie ahead for the U.S aerospace industry.

Highlights of the year in aerospace include:

Sales

Total aerospace industry sales rebounded to near record levels—rising 6%, or $8 billion, to $153 billion. DoD sales increased 4%, or $2 billion, to $49 billion, while sales to NASA and other civil agencies jumped 8%, or $1.1 billion, to $14.5 billion. Accounting for the majority of the overall growth, sales to other customers grew 6.6%, or $4 billion, to $64 billion.

Similarly, all product group categories sales grew in 2001. The second year of rising procurement (after nine straight years of decline) combined with substantial missile defense funding, helped push up missile sector sales to $10.4 billion. Both the space sector and the related products and services group saw sales increase by more than $1 billion.

Aircraft sales rose by $4.7 billion last year. Both the military and civil aircraft sectors experienced increased sales. Military sales rose $1.0 billion to $35 billion—solely on higher domestic funding, while civil sales grew $3.7 billion to $51 billion—solely due to increased exports.

For 2001, aerospace industry output accounted for 1.5% of the Gross Domestic Product and 3.9% of manufacturing sales.

Earnings

The aerospace industry generated $6.6 billion in net income after taxes on $169 billion of corporate sales last year. 2001's net, while higher than profits posted in any year before 1996, represented a decline of 36% from 1999's record-setting level. Net profit as a percentage of sales fell to 3.9%—the lowest level since 1995—down from 6.5% in 1999 and 4.7% in 2000. The corresponding profit margin for all manufacturing corporations in 2001 was 0.8%—off sharply from 2000's 6.1%. Similarly, the aerospace industry's net profit as a percentage of assets and shareholders' equity declined to 3.6% and 11.6%, respectively. For comparison, prior year aerospace returns were 4.3% and 14.8%; and the averages for all manufacturing corporations in 2001 were 0.8% and 1.9%, respectively.

Orders and Backlog

New orders fell 13% to $121 billion in 2001. While military orders rose nicely from $55 billion to $64 billion, civil orders plummeted—down 32%, or $27 billion, to $58 billion. Orders for aircraft, engines, and parts fell sharply. Combined aircraft sector orders fell $34 billion, or 35%, to $62 billion with military orders declining $9.6 billion, or 31%, to $22 billion and civil orders falling $25 billion, or 38%, to $41 billion. [Please note that the JSF award appears to be recorded outside of the aircraft sector in "Other Aerospace, Military" which jumped 263% to $26 billion.]

The industry's year-end backlog, on the other hand, grew $4.6 billion to $220 billion. Growth in the military backlog continued for its second straight year after ending a three-year slide; and the civil backlog dropped $12 billion. After reaching record levels in 2000, unfilled orders for missiles, space, and rocket propulsion receded to $33 billion in 2001. Similarly, the aircraft sector declined from an eight-year-high backlog of $156 billion for year-end 2000 to $144 billion. "Other Aerospace" unfilled orders doubled to a record $31 billion.

Civil Aircraft

The civil aircraft industry produced and delivered fewer aircraft than in 2000, but more than in any other year since 1982—3,559 in all. The decline came from general aviation aircraft, whose shipments fell by 184 to 2,612, and from civil helicopters, whose production dropped from 493 to 415. General aviation production and billings had been steadily increasing since 1994 when Congress enacted the General Aviation Revitalization Act.

Transport aircraft production, on the other hand, rose by 41 from 2000's 485. For the first time in 13 years, industry shipped more jetliners domestically than it did overseas—exporting 253 versus delivering 273 transports domestically.

According to Air BP's annual "Turbine-Engined Fleets of the World's Airlines," the world's airline fleet grew to 25,963 aircraft. There were more Boeing 737s in service than any other single type, with a total active fleet of 3,667. The McDonnell Douglas MD-80 was the next largest, numbering 1,141 aircraft.

International Civil Aviation Organization statistics show that the world's airlines carried fewer passengers and flew fewer passenger-miles in 2001. Load factors fell too from historic high levels despite airlines quickly adjusting to lower traffic by grounding aircraft.

Military Aircraft Production

Military aircraft production, as measured by acceptances, fell to its lowest level in at least 50 years. Manufacturers delivered 325 aircraft of which 46%

were destined to customers other than the U.S. government. The number of military aircraft exported declined for the fourth straight year—down 46 to 149—and those exported through the Foreign Military Sales program fell to their lowest level on record (22). Encouragingly, aircraft acceptances by the services for their own use increased for the second straight year.

U.S. military agencies accepted, for their own use, 176 aircraft in 2001—38 more than in 2000. Unfortunately, the high dollar categories, fighter/attack and transport/tanker, saw fewer acceptances. Consequently, the total flyaway cost declined for the third straight year to $7.5 billion—the lowest level since 1980.

Three programs dominated new aircraft procurement in FY 2001: the Air Force's C-17 Globemaster III cargo aircraft remained the largest, big-ticket item with 12 planes worth $3.0 billion; followed by the Navy's 39 F/A-18E/F Super Hornet fighters costing $2.8 billion and the Air Force's ten F-22 Raptor fighters costing $2.5 billion. Nine V-22 Osprey tiltrotor aircraft for $1.2 billion logged a distant third. Scheduled for significant procurement funding increases in FY 2002 are the Air Force's C-17, F-22, and unmanned aircraft and the Navy's F/A-18. Despite the number of V-22 aircraft procured increasing, V-22 procurement funding was cut 19% to $942 million.

Foreign Trade

Aerospace exports increased by $3.8 billion to $59 billion. Both civil and military products contributed to the rise. Military aircraft parts exports offset reduced military aircraft exports. However, civil transports accounted for much of the rise—up $2.5 billion to $22 billion.

The aerospace industry continued to enjoy a trade surplus even as the manufacturing sector, as a whole, saw its trade deficit hover near $400 billion. Still, the aerospace trade surplus receded for the third straight year—falling $0.7 billion to $26 billion. The all-important factor was the growth of imports to its sixth consecutive record. Imports of complete aircraft have more than doubled from 1998 levels to $15 billion. Though transport-category imports led the way, general aviation imports also accounted for significant gains—reaching $6 billion and nearly doubling in those three years.

Space Programs

A strong rebound in sales of military space vehicle systems could not fully offset non-military declines in 2001. Military space sales jumped 15% to $4.3 billion, while non-military sales fell 21%, according to data compiled by the Bureau of the Census. Sales of engines and propulsion units for missiles and space vehicles declined for both civil and military, with military space/missile propulsion dropping 44% to $382 million.

NASA estimated total federal spending for space activities increased in FY 2001 to $27 billion. DoD's and the Energy Department's space-related outlays declined while NASA's, Commerce's, and "Other" increased.

The costs to build pieces of the International Space Station and fly them to orbit aboard the Space Shuttle dominated NASA's budget. Station and Shuttle budget items together accounted for 37% of NASA's $14.3 billion in total budget authority for 2001. NASA's FY 2002 budget should rise to $14.9 billion, but the International Space Station's funding was cut 19% to $1.7 billion.

Missile Programs

Total DoD outlays for missile procurement rose for the second straight year in FY 2001 following a nine-year slide. Major programs include: Navy's Trident II with $437 million in FY 2001 funding; BMDO's Patriot, $385 million; Army/USMC Javelin, $348 million; and Army's ATACMS, $311 million.

Ballistic Missile Defense continues to dominate missile RDT&E. Funding totaled $4.2 billion in FY 2001 and is scheduled to increase to $7.0 billion in 2002.

Also, net new orders fell nearly 50% after more than doubling in 2000. The year-end 2001 backlog of unfilled orders stood at $8.4 billion.

Research and Development

R&D spending by the federal government increased $12 billion after languishing at the same real spending level for six years. The DoD, with its $45 billion in outlays, continued to be the government's largest single spender on R&D— accounting for more than half of all federal funding. NASA's R&D totaled $7 billion; and the Department of Energy was close behind with $6.6 billion. Other agencies also saw healthy growth. Organizations such as the National Science Foundation, the National Institutes of Health, and the Transportation and Agriculture Departments collectively saw their collective R&D outlays rise 19% to $27 billion. Federally-funded R&D is scheduled to increase $11 billion to $97 billion in FY 2002.

In addition to the previously mentioned Ballistic Missile Defense, the F-22 Raptor received $1.4 billion of RDT&E funding in FY 2001. Both the Joint Strike Fighter and the Army's RAH-66 Comanche are gearing up for substantially increased RDT&E funding in FY 2002-2003, as are unmanned air vehicles.

Employment

Industry employment fell to its lowest level in at least 43 years to stand at 790,000. On an annual average basis, total employment declined by 9,000. Production workers lost work at a higher rate than other categories of employees. Most of the jobs losses were in the aircraft, engines, and parts manufacturing sector.

Despite the losses, however, experts agree that engineers and technical workers remain in demand. Most aerospace companies say they still need mechanical, structural, and aerodynamics engineers, as well as software engineers and information technology specialists.

Aerospace constituted 4.5% of all manufacturing employment and 7.4% of durable goods manufacturing employment. The aerospace share of manufacturing and durable goods employment has fallen from 6.8% and 11.7% in 1990, respectively.

STANDARD INDUSTRIAL CLASSIFICATIONS
APPLICABLE TO THE AEROSPACE INDUSTRY

3721 AIRCRAFT
37211 Military aircraft
37215 Civilian aircraft
37217 Modification, conversion, and overhaul of previously accepted aircraft
37218 Aeronautical services on complete aircraft, nec

3724 AIRCRAFT ENGINES AND ENGINE PARTS
37241 Aircraft engines for military aircraft
37242 Aircraft engines for civilian aircraft
37243 Aeronautical services on aircraft engines
37244 Aircraft engine parts and accessories

3728 AIRCRAFT PARTS AND AUXILIARY EQUIPMENT, NEC
37281 Aircraft parts and auxiliary equipment, nec
37282 Aircraft propellers and helicopter rotors
37283 Research and development on aircraft parts

3761 GUIDED MISSILES AND SPACE VEHICLES
37611 Complete guided missiles (excluding propulsion systems)
37612 Complete space vehicles (excluding propulsion systems)
37613 Research and development on complete guided missiles
37614 Research and development on complete space vehicles
37615 All other services on complete guided missiles and space vehicles

3663 RADIO AND TELEVISION COMMUNICATIONS EQUIPMENT
36631 Communication systems and equipment, except broadcast

3764 SPACE PROPULSION UNITS AND PARTS
37645 Complete missile or space vehicle engines and/or propulsion units
37646 Research and development on complete missile or space vehicle engines and/or propulsion units
37647 Services on complete guided missile or space vehicle engines and/or propulsion units, nec
37648 Missile and space vehicle engine and/or propulsion unit parts and accessories

3769 SPACE VEHICLE EQUIPMENT, NEC
37692 Missile and space vehicle components, parts and subassemblies, nec
37694 Research and development on missile and space vehicle parts and components, nec

3669 COMMUNICATIONS EQUIPMENT, NEC
36691 Alarm systems
36692 Traffic control equipment
36693 Intercommunication equipment

3812 SEARCH, DETECTION, NAVIGATION, GUIDANCE, AERONAUTICAL AND NAUTICAL SYSTEMS, INSTRUMENTS, AND EQUIPMENT
38121 Aeronautical, nautical, and navigational instruments, not sending or receiving radio signals
38122 Search, detection, navigation, and guidance systems and equipment

3829 MEASURING AND CONTROLLING DEVICES, NEC
38291 Aircraft engine instruments, except flight

Source: Office of Management and Budget, "Standard Industrial Classification Manual, 1987."
NOTE: The Standard Industrial Classification (SIC) is a system developed by the U.S. Government to define the industrial composition of the economy, facilitating comparability of statistics. It is revised periodically to reflect the changing industrial composition of the economy.
NEC: Not elsewhere classified.

NORTH AMERICAN INDUSTRY CLASSIFICATION SYSTEM CODES APPLICABLE TO THE AEROSPACE INDUSTRY

33641	**Aerospace product & parts mfg**
336411	**Aircraft mfg**
3364111	Military aircraft
3364113	Civilian aircraft
3364115	Modification, conversion, and overhaul of previously accepted aircraft
3364117	Other aeronautical services on complete aircraft, nec
336411W	Aircraft manufacturing, nsk
336412	**Aircraft engine & engine parts mfg**
3364121	Military aircraft engines
3364123	Civilian aircraft engines
3364125	Aeronautical services on aircraft engines
3364127	Aircraft engine parts and accessories
336412W	Aircraft engines and engine parts manufacturing, nsk
336413	**Aircraft parts and auxiliary equipment mfg, nec**
3364131	Aircraft propellers and helicopter rotors
3364133	Research and development on aircraft parts (except engines)
3364135	Aircraft parts and auxiliary equipment, excluding hydraulic and pneumatic subassemblies and engines
336413W	Aircraft parts and auxiliary equipment, nec, nsk
336414	**Guided missile & space vehicle mfg**
3364141	Complete guided missiles
3364143	Research and development on complete guided missiles
3364145	Other services on complete guided missiles
3364147	Complete space vehicles (excluding propulsion systems)
3364149	Research and development on complete space vehicles
336414A	All other services on complete space vehicles
336414W	Guided missile and space vehicle manufacturing, nsk
336415	**Guided missile & space vehicle propulsion unit & parts mfg**
3364151	Complete missile or space vehicle engines and or propulsion units
3364153	Research and development on complete missile or space vehicle engines and or propulsion units
3364155	Other services on complete missile or space vehicle engines and or propulsion units

3364157	Missile and space vehicle engine and or propulsion parts and accessories
336415W	Space propulsion units and parts, nsk
336419	**Other guided missile & space vehicle parts & auxiliary equip mfg**
3327221	**Aircraft (including aerospace) fasteners other than plastics (meet specifications for flying vehicles)**
3345191	**Aircraft engine instruments mfg, except flight**
332912	**Fluid power valve and hose fitting mfg**
3329121	Aerospace type hydraulic fluid power valves
3329123	Aerospace type pneumatic fluid power valves
332912F	Aerospace type hydraulic and pneumatic fluid power hose or tube end fittings and assemblies
33399	**All other general purpose machinery mfg**
3339957	Aerospace type fluid power cylinders and actuators, hydraulic and pneumatic
3339967	Aerospace type fluid power pumps and motors
3339996	Filters for hydraulic and pneumatic fluid power systems, aerospace
3342201	**Communication systems and equipment, except broadcast, but including microwave equipment, and space satellites**
334290	**Alarm systems, traffic control equipment, and intercommunication and paging systems mfg**
334511	**Search, detection, navigation, guidance, aeronautical, and nautical systems and instruments mfg**
3345111	Aeronautical, nautical, and navigational instruments, not sending or receiving radio signals, except engine instruments
3345113	Search, detection, navigation, and guidance systems and equipment
334511W	Search, detection, navigation, guidance, aeronautical, and nautical systems and instruments, nsk

Source: Office of Management and Budget, "North American Industry Classification System, United States, 1997."

AEROSPACE INDUSTRY SALES BY PRODUCT GROUP
Calendar Years 1987–2001
(Millions of Dollars)

Year	TOTAL	Aircraft			Missiles	Space	Related Products & Services
		TOTAL	Civil	Military			

CURRENT DOLLARS

Year	TOTAL	TOTAL	Civil	Military	Missiles	Space	Related Products & Services
1987	$110,008	$59,188	$15,465	$43,723	$10,219	$22,266	$18,335
1988	114,562	60,886	19,019	41,867	10,270	24,312	19,094
1989	120,534	61,550	21,903	39,646	13,622	25,274	20,089
1990	134,375	71,353	31,262	40,091	14,180	26,446	22,396
1991	139,248	75,918	37,443	38,475	10,970	29,152	23,208
1992	138,591	73,905	39,897	34,008	11,757	29,831	23,099
1993	123,183	65,829	33,116	32,713	8,451	28,372	20,531
1994	110,558	57,648	25,596	32,052	7,563	26,921	18,426
1995	107,782	55,048	23,965	31,082	7,386	27,385	17,964
1996	116,812	60,296	26,869	33,427	8,008	29,040	19,469
1997	131,582	70,804	37,428	33,376	8,037	30,811	21,930
1998	147,991	83,951	49,676	34,275	7,730	31,646	24,665
1999	153,707	88,731	52,931	35,800	8,825	30,533	25,618
2000[r]	144,741	81,612	47,580	34,032	9,298	29,708	24,123
2001	153,143	86,314	51,327	34,987	10,445	30,860	25,524

CONSTANT DOLLARS[a]

Year	TOTAL	TOTAL	Civil	Military	Missiles	Space	Related Products & Services
1987	$110,008	$59,188	$15,465	$43,723	$10,219	$22,266	$18,335
1988	112,869	59,986	18,738	41,248	10,118	23,953	18,812
1989	114,250	58,341	20,761	37,579	12,912	23,956	19,042
1990	123,734	65,703	28,786	36,916	13,057	24,352	20,622
1991	124,998	68,149	33,611	34,538	9,847	26,169	20,833
1992	118,555	63,221	34,129	29,092	10,057	25,518	19,760
1993	102,482	54,766	27,551	27,215	7,031	23,604	17,081
1994	90,104	46,983	20,861	26,122	6,164	21,941	15,017
1995	86,572	44,215	19,249	24,965	5,933	21,996	14,429
1996	92,196	47,590	21,207	26,383	6,320	22,920	15,366
1997	102,959	55,402	29,286	26,116	6,289	24,109	17,160
1998	115,168	65,332	38,658	26,673	6,016	24,627	19,195
1999	118,784	68,571	40,905	27,666	6,820	23,596	19,798
2000[r]	108,583	61,224	35,694	25,530	6,975	22,287	18,097
2001	112,029	63,141	37,547	25,594	7,641	22,575	18,672

Source: Aerospace Industries Association.
NOTE: See Glossary for explanation of "Aerospace Industry," "Aerospace Sales," "Other Customers," and "Related Products and Services."
   a Based on AIA's aerospace composite deflator, 1987=100.

AEROSPACE INDUSTRY SALES BY CUSTOMER
Calendar Years 1987–2001
(Millions of Dollars)

Year	TOTAL	Aerospace Products and Services				Related Products and Services
		TOTAL	U.S. Government		Other Customers	
			Dept. of Defense	NASA and Other Agencies		

CURRENT DOLLARS

Year	TOTAL	TOTAL	Dept. of Defense	NASA and Other Agencies	Other Customers	Related Products and Services
1987	$110,008	$ 91,673	$61,817	$ 6,813	$23,043	$18,335
1988	114,562	95,468	61,327	7,899	26,242	19,094
1989	120,534	100,445	61,199	9,601	29,645	20,089
1990	134,375	111,979	60,502	11,097	40,379	22,396
1991	139,248	116,040	55,922	11,739	48,379	23,208
1992	138,591	115,493	52,202	12,408	50,882	23,099
1993	123,183	102,653	47,017	12,255	43,380	20,531
1994	110,558	92,132	43,795	11,932	36,405	18,426
1995	107,782	89,818	42,401	11,413	36,004	17,964
1996	116,812	97,344	42,535	12,391	42,418	19,469
1997	131,582	109,651	43,702	12,753	53,196	21,930
1998	147,991	123,326	42,937	13,343	67,047	24,665
1999	153,707	128,089	45,703	13,400	68,986	25,618
2000[r]	144,741	120,617	47,505	13,382	59,730	24,123
2001	153,143	127,619	49,479	14,462	63,678	25,524

CONSTANT DOLLARS [a]

Year	TOTAL	TOTAL	Dept. of Defense	NASA and Other Agencies	Other Customers	Related Products and Services
1987	$110,008	$ 91,673	$61,817	$ 6,813	$23,043	$18,335
1988	112,869	94,057	60,421	7,782	25,854	18,812
1989	114,250	95,209	58,009	9,100	28,100	19,042
1990	123,734	103,111	55,711	10,218	37,181	20,622
1991	124,998	104,165	50,199	10,538	43,428	20,833
1992	118,555	98,796	44,655	10,614	43,526	19,760
1993	102,482	85,402	39,116	10,196	36,090	17,081
1994	90,104	75,087	35,693	9,725	29,670	15,017
1995	86,572	72,143	34,057	9,167	28,919	14,429
1996	92,196	76,830	33,571	9,780	33,479	15,366
1997	102,959	85,799	34,196	9,979	41,624	17,160
1998	115,168	95,974	33,414	10,384	52,177	19,195
1999	118,784	98,987	35,319	10,355	53,312	19,798
2000[r]	108,583	90,485	35,638	10,039	44,809	18,097
2001	112,029	93,357	36,195	10,579	46,582	18,672

Source: Aerospace Industries Association.
NOTE: See Glossary for explanation of "Aerospace Industry," "Aerospace Sales," "Other Customers," and "Related Products and Services."
a Based on AIA's aerospace composite price deflator, 1987=100.

SALES OF MAJOR AEROSPACE COMPANIES
AS REPORTED BY THE BUREAU OF THE CENSUS
Calendar Years 1987–2001
(Millions of Dollars)

Year	GRAND TOTAL	TOTAL Military	TOTAL Non-Mil.	Aircraft, Engines, & Parts Military	Aircraft, Engines, & Parts Non-Mil.	Missiles, Space, & Rocket Propulsion	Other Aerospace Military	Other Aerospace Non-Mil.	Non-Aerospace
CURRENT DOLLARS									
1987	$110,301	$70,194	$40,107	$27,806	$21,256	$20,715	$15,786	$3,429	$21,309
1988	113,548	69,448	44,100	25,068	25,674	21,514	16,382	2,946	21,964
1989	122,148	71,647	50,501	24,287	29,538	22,643	16,908	3,605	25,167
1990	136,646	73,616	63,030	27,667	38,622	22,040	15,773	4,342	28,202
1991	123,862	67,089	56,773	25,385	43,155	23,311	13,472	4,281	14,258
1992	118,736	61,410	57,326	23,509	44,160	21,349	12,153	3,377	14,188
1993	109,926	56,102	53,824	20,099	40,987	18,134	11,936	3,592	15,178
1994	104,296	58,012	46,284	23,652	30,901	18,406	11,981	4,417	14,939
1995	102,797	52,476	50,321	22,944	32,085	18,366	11,921	4,462	13,019
1996	103,115	53,153	49,962	24,804	32,722	18,506	12,171	4,624	10,287
1997	114,946	50,648	64,298	23,944	42,614	21,354	12,320	3,922	10,792
1998	119,258	45,110	74,148	23,795	52,708	16,109	7,818	5,035	13,796
1999	124,181	49,690	74,491	26,043	56,406	15,661	9,062	4,472	12,535
2000 r	109,311	43,256	66,055	23,196	46,477	15,603	6,035	4,785	13,215
2001	117,343	47,230	70,113	22,127	52,768	15,502	8,204	5,731	13,011
CONSTANT DOLLARS [a]									
1987	$110,301	$70,194	$40,107	$27,806	$21,256	$20,715	$15,786	$3,429	$21,309
1988	111,870	68,422	43,448	24,698	25,295	21,196	16,140	2,902	21,639
1989	115,780	67,912	47,868	23,021	27,998	21,463	16,027	3,417	23,855
1990	125,825	67,786	58,039	25,476	35,564	20,295	14,524	3,998	25,969
1991	111,187	60,224	50,963	22,787	38,739	20,925	12,093	3,843	12,799
1992	101,571	52,532	49,038	20,110	37,776	18,263	10,396	2,889	12,137
1993	91,453	46,674	44,779	16,721	34,099	15,087	9,930	2,988	12,627
1994	85,001	47,280	37,721	19,276	25,184	15,001	9,764	3,600	12,175
1995	82,568	42,149	40,418	18,429	25,771	14,752	9,575	3,584	10,457
1996	81,385	41,952	39,433	19,577	25,826	14,606	9,606	3,650	8,119
1997	89,942	39,631	50,311	18,736	33,344	16,709	9,640	3,069	8,444
1998	92,808	35,105	57,703	18,518	41,018	12,536	6,084	3,918	10,736
1999	95,967	38,400	57,566	20,126	43,590	12,103	7,003	3,456	9,687
2000 r	82,004	32,450	49,554	17,401	34,866	11,705	4,527	3,590	9,914
2001	85,840	34,550	51,290	16,187	38,601	11,340	6,001	4,192	9,518

Source: Bureau of the Census, "Aerospace Industry (Orders, Sales, and Backlog)" (Annually).
   a   Based on AIA's aerospace composite price deflator, 1987=100.

ORDERS AND BACKLOG OF MAJOR AEROSPACE COMPANIES
AS REPORTED BY THE BUREAU OF THE CENSUS
Calendar Years 1987–2001
(Millions of Dollars)

Year	GRAND TOTAL	TOTAL Military	TOTAL Non-Mil.	Aircraft, Engines, & Parts Military	Aircraft, Engines, & Parts Non-Mil.	Missiles, Space, & Rocket Propulsion	Other Aerospace Military	Other Aerospace Non-Mil.	Non-Aerospace
NET NEW ORDERS									
1987	$121,224	$ 67,594	$ 53,630	$19,347	$ 33,000	$26,272	$14,178	$4,379	$24,048
1988	147,128	69,209	77,919	24,242	57,906	20,240	18,423	3,044	23,273
1989	173,635	79,992	93,643	28,818	67,773	26,820	17,814	3,945	28,465
1990	145,965	56,405	89,560	17,735	64,651	20,207	12,945	3,556	26,871
1991	122,485	63,017	59,468	26,675	40,815	24,955	11,329	4,360	14,351
1992	100,306	57,383	42,923	19,631	30,110	22,849	11,201	3,256	13,259
1993	79,770	49,541	30,229	19,518	16,090	14,919	11,121	4,629	13,493
1994	88,706	53,268	35,438	23,352	20,166	13,705	12,924	5,395	13,164
1995	109,109	49,350	59,759	19,854	36,467	19,181	13,716	5,261	14,630
1996	126,267	62,127	64,140	25,343	45,281	27,067	12,136	5,070	11,370
1997	118,993	47,802	71,192	21,424	49,676	21,326	12,348	4,125	10,096
1998	109,993	38,678	71,314	16,870	47,613	19,699	7,628	4,468	13,715
1999	115,257	49,696	65,561	25,009	48,018	18,824	10,261	4,152	8,992
2000 [r]	140,086	54,525	85,165	31,396	65,459	18,368	7,046	3,900	13,917
2001	121,395	63,503	57,892	21,752	40,746	12,850	25,557	5,042	15,448
BACKLOG AS OF DECEMBER 31									
1987	$158,650	$ 99,474	$ 59,176	$36,514	$ 43,501	$30,544	$18,937	$4,604	$24,550
1988	191,518	99,117	92,401	35,515	75,765	29,078	20,584	4,734	25,842
1989	252,401	114,070	138,331	44,026	115,124	33,771	24,186	7,652	27,642
1990	250,079	88,471	161,608	33,788	139,152	31,648	18,501	4,999	21,991
1991	245,241	89,517	155,724	39,149	134,527	32,657	17,213	4,907	16,788
1992	236,076	92,139	143,937	44,255	124,322	32,933	14,886	4,859	14,821
1993	211,814	91,751	120,063	46,177	96,228	29,511	16,668	7,958	15,272
1994	192,561	84,445	108,116	44,624	85,305	24,746	15,599	8,043	14,244
1995	202,638	82,309	120,329	44,642	92,239	27,113	17,534	8,214	12,906
1996	229,871	89,500	140,371	47,635	106,341	35,440	16,176	9,339	14,940
1997	218,951	78,870	140,082	43,615	111,931	34,585	12,125	4,754	11,942
1998	200,288	69,962	130,326	37,530	106,166	31,174	9,665	3,488	12,264
1999	188,409	68,379	120,029	36,565	96,596	33,880	9,904	3,051	8,413
2000 [r]	214,966	73,741	141,225	41,250	115,241	36,283	10,028	4,081	8,083
2001	219,556	90,128	129,428	40,941	102,809	33,268	27,395	3,382	11,761

Source: Bureau of the Census, "Aerospace Industry (Orders, Sales, and Backlog)" (Annually).

15

AEROSPACE SALES AND THE NATIONAL ECONOMY
Calendar Years 1987–2001
(Billions of Dollars)

Year	Gross Domestic Product	Industry Sales			Aerospace Sales as Percent of		
		Manufacturing[r]	Durable Goods[r]	Aero-space	GDP	Manufacturing[r]	Durable Goods[r]
CURRENT DOLLARS							
1987	$ 4,742.5	$2,474.0	$1,296.2	$110.0	2.3%	4.4%	8.5%
1988	5,108.3	2,694.0	1,420.5	114.6	2.2	4.3	8.1
1989	5,489.1	2,837.8	1,475.9	120.5	2.2	4.2	8.2
1990	5,803.2	2,909.0	1,483.1	134.4	2.3	4.6	9.1
1991	5,986.2	2,878.8	1,453.2	139.2	2.3	4.8	9.6
1992	6,318.9	2,872.0	1,486.8	138.6	2.2	4.8	9.3
1993	6,642.3	2,985.5	1,569.5	123.2	1.9	4.1	7.8
1994	7,054.3	3,191.0	1,717.0	110.6	1.6	3.5	6.4
1995	7,400.5	3,414.0	1,837.1	107.8	1.5	3.2	5.9
1996	7,813.2	3,526.0	1,907.4	116.8	1.5	3.3	6.1
1997	8,318.4	3,756.2	2,068.9	131.6	1.6	3.5	6.4
1998	8,781.5	3,824.8	2,156.5	148.0	1.7	3.9	6.9
1999[r]	9,274.3	3,948.4	2,243.2	153.7	1.7	3.9	6.9
2000[r]	9,824.6	4,124.5	2,286.0	144.7	1.5	3.5	6.3
2001	10,082.2	3,897.7	2,100.1	153.1	1.5	3.9	7.3

Year					Real Annual Growth[b]			
CONSTANT DOLLARS[a]					GDP	Mfg.[r]	Durs.[r]	Aero.
1987	$6,111.5	$3,188.1	$1,670.3	$110.0	3.3%	2.8%	1.6%	3.4%
1988	6,369.5	3,359.1	1,771.1	112.9	4.2	5.4	6.0	2.6
1989	6,589.6	3,406.7	1,771.8	114.3	3.5	1.4	0.0	1.2
1990	6,708.9	3,363.0	1,714.5	123.7	1.8	(1.3)	(3.2)	8.3
1991	6,673.6	3,209.3	1,620.1	125.0	(0.5)	(4.6)	(5.5)	1.0
1992	6,883.3	3,128.5	1,619.6	118.6	3.1	(2.5)	(0.0)	(5.2)
1993	7,058.8	3,172.7	1,667.9	102.5	2.5	1.4	3.0	(13.6)
1994	7,348.2	3,324.0	1,788.5	90.1	4.1	4.8	7.2	(12.1)
1995	7,543.8	3,480.1	1,872.7	86.6	2.7	4.7	4.7	(3.9)
1996	7,813.2	3,526.0	1,907.4	92.2	3.6	1.3	1.9	6.5
1997	8,163.3	3,686.2	2,030.3	103.0	4.5	4.5	6.4	11.7
1998	8,509.2	3,706.2	2,089.7	115.2	4.2	0.5	2.9	11.9
1999[r]	8,858.0	3,771.1	2,142.5	118.8	4.1	1.8	2.5	3.1
2000[r]	9,181.9	3,854.7	2,136.5	108.6	3.7	2.2	(0.3)	(8.6)
2001	9,215.9	3,562.8	1,919.7	112.0	0.4	(7.6)	(10.1)	3.2

Source: Council of Economic Advisors, "Economic Indicators" (Monthly); Bureau of Census; and Aerospace Industries Association.
 a Aerospace industry constant dollar sales based on AIA's aerospace composite price deflator, 1987=100. Others based on GDP deflator, 1996=100.
 b Parentheses indicate negative real annual growth.

GROSS DOMESTIC PRODUCT, FEDERAL BUDGET, AND DEFENSE BUDGET

Fiscal Years 1970–2003
(Billions of Dollars)

Year	Fiscal Year GDP	Federal Budget Outlays		Defense Outlays as Percent of	
		Net Total[a]	National Defense[b]	GDP	Federal Budget
1970	$ 1,013.2	$ 195.6	$ 81.7	8.1%	41.8%
1971	1,081.4	210.2	78.9	7.3	37.5
1972	1,181.5	230.7	79.2	6.7	34.3
1973	1,308.1	245.7	76.7	5.9	31.2
1974	1,442.1	269.4	79.3	5.5	29.5
1975	1,559.8	332.3	86.5	5.5	26.0
1976	1,736.7	371.8	89.6	5.2	24.1
Tr.Qtr.	454.8	96.0	22.3	4.9	23.2
1977	1,971.3	409.2	97.2	4.9	23.8
1978	2,218.6	458.7	104.5	4.7	22.8
1979	2,503.8	504.0	116.3	4.6	23.1
1980	2,732.1	590.9	134.0	4.9	22.7
1981	3,061.6	678.2	157.5	5.1	23.2
1982	3,228.6	745.8	185.3	5.7	24.8
1983	3,440.5	808.4	209.9	6.1	26.0
1984	3,839.4	851.9	227.4	5.9	26.7
1985	4,136.6	946.4	252.7[b]	6.1	26.7
1986	4,401.4	990.5	273.4	6.2	27.6
1987	4,647.0	1,004.1	282.0	6.1	28.1
1988	5,014.7	1,064.5	290.4	5.8	27.3
1989	5,405.5	1,143.7	303.6	5.6	26.5
1990	5,735.6	1,253.2	299.3	5.2	23.9
1991	5,930.4	1,324.4	273.3[c]	4.6	20.6
1992	6,218.6	1,381.7	298.4[c]	4.8	21.6
1993	6,558.4	1,409.5	291.1[c]	4.4	20.7
1994	6,944.6	1,461.9	281.6	4.1	19.3
1995	7,324.0	1,515.8	272.1	3.7	17.9
1996	7,694.6	1,560.6	265.8	3.5	17.0
1997	8,185.2	1,601.3	270.5	3.3	16.9
1998	8,663.9[r]	1,652.6	268.5	3.1	16.2
1999[r]	9,124.3	1,701.9	274.9	3.0	16.2
2000	9,744.3[r]	1,788.8	294.5	3.0	16.5
2001	10,150.5	1,863.9	308.5	3.0	16.6
2002[E]	10,361.6	2,052.3	348.0	3.4	17.0
2003[E]	10,922.3	2,128.2	379.0	3.5	17.8

Source: Office of Management and Budget, "The Budget of the United States Government" (Annually).

a "Net Total" is government-wide total less intragovernmental transactions.

b "National Defense" includes the military budget of DoD and other defense-related activities. Beginning in 1985, the Federal Budget reflects establishment of a military retirement trust fund. Data for prior years adjusted for comparable treatment of military retired pay.

c 1991–1993 reflects transfers from the Defense Cooperation Account funded by foreign government and private cash contributions reducing total U.S.-funded military outlays.

17

FEDERAL OUTLAYS FOR DEFENSE, NASA, AND AEROSPACE PRODUCTS AND SERVICES

Fiscal Years 1975–2003
(Millions of Dollars)

Year	Total National Defense	Total NASA	Federal Outlays for Aerospace Products and Services			Aerospace as Percent of Total National Defense and NASA
			TOTAL	DoD[a]	NASA	
1975	$ 86,509	$ 3,267	$11,544	$ 8,373	$ 3,181	12.9%
1976	89,619	3,669	12,364	8,816	3,548	13.3
Tr.Qtr.	22,269	951	2,855	1,959	926	12.3
1977	97,241	3,945	13,229	9,389	3,840	13.1
1978	104,495	3,983	13,926	10,067	3,859	12.8
1979	116,342	4,197	16,686	12,622	4,064	13.8
1980	133,995	4,852	20,269	15,558	4,711	14.6
1981	157,513	5,421	24,276	19,002	5,274	14.9
1982	185,309	6,035	29,501	23,575	5,926	15.4
1983	209,903	6,664	35,364	28,808	6,556	16.3
1984	227,413	7,048	39,663	32,723	6,940	16.9
1985	252,748	7,318	44,483	37,335	7,148	17.1
1986	273,375	7,404	49,773	42,558	7,215	17.7
1987	281,999	7,591	51,871	44,429	7,442	17.9
1988	290,361	9,092	48,848	39,922	8,926	16.3
1989	303,559	11,036	52,933	42,072	10,861	16.8
1990	299,331	12,429	53,194	40,992	12,202	17.1
1991[b]	273,292	13,878	53,630	40,089	13,541	18.7
1992[b]	298,350	13,961	50,569	37,085	13,484	16.2
1993[b]	291,086	14,305	45,496	31,763	13,733	14.9
1994	281,642	13,695	41,082	27,774	13,308	13.9
1995	272,066	13,378	36,696	23,638	13,058	12.9
1996	265,753	13,881	32,947	20,530	12,417	11.8
1997	270,505	14,360	32,808	19,888	12,920	11.5
1998	268,456	14,206	33,184	20,380	12,804	11.7
1999	274,873	13,664	32,968	20,564	12,404	11.4
2000	294,495	13,442	34,617	22,222	12,395	11.2
2001	308,533	14,199	36,721	23,420	13,301	11.4
2002[E]	347,986	14,484	38,383	24,419	13,964	10.6
2003[E]	379,012	14,885	39,462	25,412	14,050	10.0

Source: Office of Management and Budget, "The Budget of the United States Government" (Annually).

NOTE: "National Defense" includes the military budget of the Department of Defense and other defense-related activities. "Total NASA" includes all categories of the NASA budget; NASA construction is not included in "Aerospace Products and Services." See additional explanation with following table.

a Outlays for aircraft and missile procurement. Does not include RDT&E, which DoD has not reported by product group since 1977, and which, for comparability, has been subtracted from data previously reported in this table for earlier years. Also included are revisions to missile procurement data.

b 1991–1993 reflects transfers from the Defense Cooperation Account funded by foreign government and private cash contributions reducing total U.S.-funded military outlays.

FEDERAL OUTLAYS FOR AEROSPACE PRODUCTS AND SERVICES

Fiscal Years 1970–2003
(Millions of Dollars)

Year	TOTAL	Department of Defense[a]			NASA[b]
		TOTAL	Aircraft	Missiles	
1970	$14,559	$10,860	$ 7,948	$ 2,912	$ 3,699
1971	12,918	9,580	6,549	3,031	3,338
1972	12,309	8,936	5,927	3,009	3,373
1973	11,360	8,089	5,066	3,023	3,271
1974	11,168	7,987	5,006	2,981	3,181
1975	11,554	8,373	5,484	2,889	3,181
1976	12,364	8,816	6,520	2,296	3,548
Tr.Qtr.	2,885	1,959	1,557	402	926
1977	13,229	9,389	6,608	2,781	3,840
1978	13,926	10,067	6,971	3,096	3,859
1979	16,686	12,622	8,836	3,786	4,064
1980	20,269	15,558	11,124	4,434	4,711
1981	24,276	19,002	13,193	5,809	5,274
1982	29,501	23,575	16,793	6,782	5,926
1983	35,364	28,808	21,013	7,795	6,556
1984	39,663	32,723	23,196	9,527	6,940
1985	44,483	37,335	26,586	10,749	7,148
1986	49,773	42,558	30,828	11,730	7,215
1987	51,871	44,429	32,956	11,473 [c]	7,442
1988	48,848	39,922	28,246	11,676	8,926
1989	52,933	42,072	27,569	14,503	10,861
1990	53,194	40,992	26,142	14,851	12,202
1991	53,630	40,089	25,689	14,400	13,541
1992	50,569	37,085	23,581	13,504	13,484
1993	45,496	31,763	20,359	11,404	13,733
1994	41,082	27,774	18,840	8,934	13,308
1995	36,696	23,638	16,125	7,513	13,058
1996	32,947	20,530	14,331	6,199	12,417
1997	32,808	19,888	14,663	5,225	12,920
1998	33,184	20,380	15,473	4,907	12,804
1999	32,968	20,564	16,484	4,080	12,404
2000	34,617	22,222	17,991	4,231	12,395
2001	36,721	23,420	17,979	5,441	13,301
2002 [E]	38,383	24,419	19,496	4,923	13,964
2003 [E]	39,462	25,412	20,023	5,389	14,050

Source: Office of Management and Budget, "The Budget of the United States Government" (Annually).
  a Outlays for aircraft and missile procurement. Does not include RDT&E, which DoD has not reported by product group since 1977, and which for comparability, has been subtracted from data previously reported in this table for earlier years.
  b Excludes Construction of Facilities, Office of Inspector General, and Air Transportation.
  c Beginning in 1978, DoD combined Navy Missile Procurement with torpedoes and other related products into Navy Weapons Procurement, of which missiles comprise approximately 80 percent.

DEPARTMENT OF DEFENSE
MILITARY OUTLAYS BY FUNCTIONAL TITLE[a]
Fiscal Years 1994–2003
(Millions of Dollars)

	1994	1995	1996	1997
TOTAL	$268,622	$259,442	$253,187	$258,311
Procurement—TOTAL	$ 61,769	$ 54,982	$ 48,913	$ 47,690
Aircraft	18,840	16,125	14,331	14,663
Missiles[b]	8,934	7,513	6,199	5,225
Ships	9,132	8,780	7,346	7,085
Weapons[b]	1,795	1,783	1,788	1,918
Ammunition	997	1,339	1,232	1,615
Other[c]	22,071	19,441	18,017	17,184
Military Personnel—TOTAL	73,137	70,809	66,669	69,724
Active Forces	63,686	61,606	57,843	60,371
Reserve Forces	9,449	9,203	8,826	9,353
RDT&E	34,762	34,594	36,494	37,015
Operations & Maintenance	87,929	91,078	88,759	92,461
Military Construction	4,979	6,823	6,683	6,187
Family Housing	3,316	3,571	3,828	4,003
Other	2,729	(2,415)	1,841	1,231

(Continued on next page)

DEPARTMENT OF DEFENSE
MILITARY OUTLAYS BY FUNCTIONAL TITLE[a]
Fiscal Years 1994–2003, continued
(Millions of Dollars)

1998	1999	2000	2001	2002 [E]	2003 [E]
$256,122	$261,380	$281,223	$293,995	$330,553	$360,989
$ 48,206	$ 48,826	$ 51,696	$ 54,986	$ 59,618	$ 62,274
15,473	16,484	17,991	17,979	19,496	20,023
4,907	4,080	4,231	5,441	4,923	5,389
6,784	6,697	6,679	7,115	7,682	8,318
1,824	1,885	1,756	1,856	2,469	2,608
1,761	1,998	1,836	2,153	2,312	2,496
17,457	17,682	19,203	20,442	22,736	23,440
68,976	69,503	75,950	73,977	81,155	92,817
59,793	59,718	65,535	63,109	69,723	78,836
9,183	9,785	10,415	10,868	11,432	13,981
37,420	37,363	37,606	40,599	45,057	50,823
93,473	96,418	105,870	113,985	133,566	143,471
6,044	5,521	5,109	5,010	5,727	5,987
3,871	3,692	3,413	3,519	3,760	3,894
(1,868)	57	1,579	1,919	1,670	1,723

Source: Office of Management and Budget, "The Budget of the United States Government" (Annually).
NOTE: Data in parentheses are credit items.
   a Includes all items in the DoD military budget; excludes the DoD civil budget for the Army Corps of Engineers and other non-defense related activites.
   b Beginning in 1978, DoD combined Navy Missiles Procurement with torpedoes and other related products into Navy Weapons Procurement. Missiles comprise approximately 80 percent of the value of this category.
   c Includes Communications and Electronics.

FEDERAL PRICE DEFLATORS FOR GDP, DEFENSE, PPI, AND CPI
Calendar/Fiscal Years 1973–2003

Year	GDP		Federal Government Defense Purchases		PPI, Capital Equipment	CPI, (Urban) All items
	FY GDP	CY GDP	Goods & Services	Equipment Investment		
	(FY 1996 =100)	(CY 1996 =100)	(CY 1996 =100)	(CY 1996 =100)	(CY 1982 =100)	(CY 82–84 =100)
1973	33.2	33.6	32.9	NA	44.2	44.4
1974	35.8	36.6	35.8	NA	50.5	49.3
1975	39.5	40.0	39.2	NA	58.2	53.8
1976	41.9	42.3	42.0	NA	62.1	56.9
1977	44.5	45.0	45.2	NA	66.1	60.6
1978	47.6	48.2	48.3	NA	71.3	65.2
1979	51.4	52.2	52.2	NA	77.5	72.6
1980	56.0	57.1	57.9	NA	85.8	82.4
1981	61.4	62.4	63.7	NA	94.6	90.9
1982	65.7	66.3	68.4	NA	100.0	96.5
1983	68.6	68.9	70.9	NA	102.8	99.6
1984	71.1	71.4	76.0	NA	105.2	103.9
1985	73.5	73.7	77.2	NA	107.5	107.6
1986	75.3	75.3	77.3	NA	109.7	109.6
1987	77.3	77.6	78.0	89.5	111.7	113.6
1988	79.9	80.2	79.7	88.8	114.3	118.3
1989	82.9	83.3	81.9	89.9	118.8	124.0
1990	86.0	86.5	84.6	91.4	122.9	130.7
1991	89.4	89.7	87.7	93.4	126.7	136.2
1992	91.7	91.8	90.8	93.9	129.1	140.3
1993	93.9	94.1	92.5	95.9	131.4	144.5
1994	96.0	96.0	94.5	98.1	134.1	148.2
1995	98.0	98.1	96.9	100.2	136.7	152.4
1996	100.0	100.0	100.0	100.0	138.3	156.9
1997	102.0	101.9	101.4	97.9	138.2	160.5
1998	103.4	103.2	102.2	95.8	137.6	163.0
1999	104.7	104.7	104.5 r	96.0	137.6	166.6
2000	106.9	107.0	107.5 r	95.7 r	138.8	172.2
2001	109.4	109.4	109.3	94.4	139.7	177.1
2002 E	111.8	111.7	NA	NA	NA	180.5
2003 E	113.8	113.7	NA	NA	NA	184.5

Source: Bureau of Economic Analysis, Price Measurement Branch; Bureau of Labor Statistics; and Office of Management and Budget, "The Budget of the United States Government" (Annually).

Key: PPI = Producer Price Index for Capital Equipment.
CPI = Consumer Price Index, All Items, All Urban Consumers for 1978 and subsequent years. Previous years, All Urban Wage Earners.
GDP = Gross Domestic Product.

PRICE DEFLATORS FOR AEROSPACE INDUSTRY
Calendar Years 1972–2001

Year	Aerospace Deflators (1987 = 100)					
	Composite	SIC 3721	SIC 3724	SIC 3728	SIC 3761	SIC 3764,9
1972	33.7	39.9	30.1	36.6	39.7	34.4
1973	37.7	41.2	30.9	38.1	39.4	35.6
1974	41.5	44.8	34.9	44.0	41.6	40.5
1975	46.6	48.3	42.3	51.6	45.2	49.2
1976	51.0	52.8	45.9	56.5	50.4	53.8
1977	54.6	56.2	49.1	58.7	55.6	58.2
1978	57.5	59.3	54.6	55.2	60.7	63.6
1979	63.5	65.3	60.9	58.9	69.7	70.0
1980	70.6	72.9	66.3	65.3	78.9	78.5
1981	79.5	80.8	77.0	74.9	87.1	89.5
1982	87.9	89.8	85.2	84.3	93.4	97.2
1983	92.2	94.4	89.5	87.9	98.6	101.5
1984	99.8	105.9	98.1	93.6	100.7	102.9
1985 [a]	98.7	100.7	99.2	94.4	102.4	103.2
1986	99.8	100.6	99.3	97.9	103.5	102.4
1987	100.0	100.0	100.0	100.0	100.0	100.0
1988	101.5	102.2	103.0	103.5	98.6	98.4
1989	105.5	111.0	105.8	106.8	97.1	97.7
1990	108.6	116.8	111.7	109.8	93.4	98.2
1991	111.4	121.3	117.0	113.6	90.5	101.9
1992 [b]	116.9	125.2	122.7	118.0	88.4	104.8
1993	120.2	129.5	124.7	120.9	90.9	109.6
1994	122.7	133.9	128.0	123.5	88.0	107.1
1995	124.5	138.3	129.9	124.4	84.3	104.2
1996	126.7	141.5	132.4	128.8	81.8	103.6
1997	127.8	143.4	133.7	131.4	77.8	103.5
1998	128.5	143.8	134.7	133.0	78.3	102.4
1999	129.4	145.1	135.7	134.3	76.8 [r]	103.1 [r]
2000 [r]	133.3	151.6	138.6	135.4	78.1	102.8
2001	136.7	156.9	142.8	138.5	76.4	102.5

Source: Aerospace Industries Association, based on data from: Bureau of Labor Statistics, Producer Price Indices; Bureau of Economic Analysis, Chain-Type Price Indexes and Implicit Price Deflators; and International Trade Administration.

a   The International Trade Administration has discontinued its reporting of the Aerospace Deflators with 1986. Subsequent composite deflators computed by AIA and deflators for 1985 and 1986 revised for consistency.

b   The Bureau of Economic Analysis discontinued its reporting in 1995 of the National Defense Purchases Deflators (used in AIA's Composite calculations). 1992-1994 revised using 1992 fixed weights and BEA's Chain-Type Price Indexes for National Defense Investment and Consumption Expenditures.

Key: SIC = Standard Industrial Classification, SIC 3721 = Aircraft; SIC 3724 = Aircraft Engines and Engine Parts; SIC 3728 = Aircraft Parts; SIC 3761 = Missiles and Space Vehicles; SIC 3764 = Space Propulsion; SIC 3769 = Space Equipment not elsewhere classified.

AIRCRAFT PRODUCTION

Sales of new aircraft, engines, and parts rose $5.2 billion from 2000's level to $75 billion, led by gains in the civil sector. Data compiled by the U.S. Census Bureau showed sales of military aircraft, engines, and parts slipped about 4.6% to $22 billion, while civil aircraft, engines, and parts sales rebounded sharply—up 14% to $53 billion following a one-year decline. Hidden in the military aircraft sector decline was a small, but strong gain in sales of military engines and parts—up $411 million from year-earlier levels to $4.0 billion.

Military aircraft production, as measured by acceptances, fell to its lowest level in at least 50 years. Manufacturers delivered 325 aircraft of which 46% were destined to customers other than the U.S. government. The number of military aircraft exported declined for the fourth straight year—down 46, or 24%, to 149—and those exported through the FMS program fell to their lowest level on record (22). Encouragingly, aircraft acceptances by the services for their own use increased for the second straight year.

Aircraft acceptances (including FMS) increased by 18 to 198. All categories showed gains except fighter/attack and tanker/transport. Corresponding gains occurred in associated flyaway costs, but the declines in fighter/attack and tanker/transport categories pulled total flyaway costs down $860 million to $7.5 billion.

The civil aircraft industry produced and delivered 221 fewer aircraft in 2001 than it had in 2000. The decline came from general aviation aircraft, whose shipments dropped by 184 to 2,618, and from civil helicopters, whose production subsided from nearly 500 to 415. The weak economy and at least partially-satiated demand in the fractional sector ended a six-year string of rising shipments and billings. Transport aircraft production, on the other hand, rebounded by 41 to 526. For the first time in 13 years, industry shipped more jetliners domestically than it did overseas—exporting 253 versus delivering

273 transports domestically. As a result, even though shipments in 2001 were 221 aircraft behind 2000, they were worth 10% more, with a total value of $42 billion.

Net new orders in 2001 fell by a third from year-earlier levels. Only the civil engine sector registered a gain—up $2.1 billion. The aircraft sector's unfilled order backlog fell $13 billion from an eight-year high of $156 billion. While both military and civil registered declining backlogs, civil was responsible for the majority of the drop—down $12 billion.

Three programs dominated new aircraft procurement in FY 2001: the Air Force's C-17 Globemaster III cargo aircraft remained the largest, big-ticket item with 12 planes worth $3.0 billion, followed by the Navy's 39 F/A-18E/F Super Hornet fighters costing $2.8 billion, and the Air Force's ten F-22 Raptor fighters costing $2.5 billion. Nine V-22 Osprey tiltrotor aircraft for $1.2 billion logged a distant fourth. Scheduled for significant procurement funding increases in FY 2002 are the Air Force's C-17, F-22, and unmanned aircraft and the Navy's F/A-18. Despite the number of V-22 aircraft increasing, V-22 procurement funding is scheduled to fall 19% to $942 million.

SALES OF AIRCRAFT, ENGINES, AND PARTS
Calendar Years 1987–2001
(Millions of Dollars)

Year	GRAND TOTAL	TOTAL		Complete Aircraft & Parts		Aircraft Engines & Parts	
		Mili-tary	Non-Mil.	Mili-tary	Non-Mil.	Mili-tary	Non-Mil.

CURRENT DOLLARS

1987	$49,062	$27,806	$21,256	$22,168	$14,862	$5,638	$ 6,394
1988	50,742	25,068	25,674	19,030	16,681	6,038	8,993
1989	53,825	24,287	29,538	18,256	20,140	6,031	9,398
1990	66,289	27,667	38,622	22,023	27,872	5,644	10,750
1991	68,540	25,385	43,155	19,710	33,215	5,675	9,940
1992	67,669	23,509	44,160	18,411	35,595	5,098	8,565
1993	61,086	20,099	40,987	16,118	32,780	3,981	8,207
1994	54,553	23,652	30,901	20,127	23,176	3,525	7,725
1995	55,029	22,944	32,085	19,596	22,897	3,348	9,188
1996	57,526	24,804	32,722	20,822	20,993	3,982	11,729
1997	66,558	23,944	42,614	21,297	33,206	2,647	9,408
1998	76,503	23,795	52,708	21,154	42,541	2,641	10,167
1999	82,449	26,043	56,406	22,917	45,107	3,126	11,299
2000[r]	69,673	23,196	46,477	19,650	37,538	3,546	8,939
2001	74,895	22,127	52,768	18,170	40,812	3,957	11,956

CONSTANT DOLLARS[a]

1987	$49,062	$27,806	$21,256	$22,168	$14,862	$5,638	$ 6,394
1988	49,992	24,698	25,295	18,749	16,434	5,949	8,860
1989	51,019	23,021	27,998	17,304	19,090	5,717	8,908
1990	61,040	25,476	35,564	20,279	25,665	5,197	9,899
1991	61,526	22,787	38,739	17,693	29,816	5,094	8,923
1992	57,886	20,110	37,776	15,749	30,449	4,361	7,327
1993	50,820	16,721	34,099	13,409	27,271	3,312	6,828
1994	44,460	19,276	25,184	16,403	18,888	2,873	6,296
1995	44,200	18,429	25,771	15,740	18,391	2,689	7,380
1996	45,403	19,577	25,826	16,434	16,569	3,143	9,257
1997	52,080	18,736	33,344	16,664	25,983	2,071	7,362
1998	59,535	18,518	41,018	16,462	33,106	2,055	7,912
1999	63,716	20,126	43,590	17,710	34,859	2,416	8,732
2000[r]	52,268	17,401	34,866	14,741	28,161	2,660	6,706
2001	54,788	16,187	38,601	13,292	29,855	2,895	8,746

Source: Bureau of the Census, "Aerospace Industry (Orders, Sales, and Backlog)" (Annually).
a Based on AIA's aerospace composite price deflator, 1987=100.

ORDERS AND BACKLOG OF AIRCRAFT, ENGINES, AND PARTS
Calendar Years 1987–2001
(Millions of Dollars)

Year	GRAND TOTAL	TOTAL		Complete Aircraft & Parts		Aircraft Engines & Parts	
		Mili-tary	Non-Mil.	Mili-tary	Non-Mil.	Mili-tary	Non-Mil.
NET NEW ORDERS							
1987	$ 52,347	$19,347	$ 33,000	$15,070	$ 24,083	$4,277	$ 8,917
1988	82,148	24,242	57,906	17,493	41,762	6,749	16,144
1989	96,591	28,818	67,773	23,569	52,619	5,249	15,154
1990	82,386	17,735	64,651	12,766	52,371	4,969	12,280
1991	67,490	26,675	40,815	22,140	30,745	4,535	10,070
1992	49,741	19,631	30,110	16,391	20,548	3,240	9,562
1993	35,608	19,518	16,090	15,853	11,238	3,665	4,852
1994	43,518	23,352	20,166	19,806	12,854	3,546	7,312
1995	56,321	19,854	36,467	16,248	27,156	3,606	9,311
1996	70,624	25,343	45,281	21,755	33,802	3,588	11,479
1997	71,100	21,424	49,676	19,102	41,439	2,322	8,237
1998	64,483	16,870	47,613	14,051	37,362	2,819	10,251
1999	73,027	25,009	48,018	21,422	35,529	3,587	12,489
2000 ʳ	96,855	31,396	65,459	27,440	54,335	3,956	11,124
2001	62,498	21,752	40,746	18,134	27,540	3,618	13,206
BACKLOG AS OF DECEMBER 31							
1987	$ 80,015	$36,514	$ 43,501	$29,869	$ 34,625	$6,645	$ 8,876
1988	111,280	35,515	75,765	28,186	59,679	7,329	16,086
1989	159,150	44,026	115,124	36,888	95,108	7,138	20,016
1990	172,940	33,788	139,152	27,259	119,123	6,529	20,029
1991	173,676	39,149	134,527	32,795	116,139	6,354	18,388
1992	168,577	44,255	124,322	39,748	107,686	4,507	16,636
1993	142,405	46,177	96,228	41,732	82,772	4,445	13,456
1994	129,929	44,624	85,305	40,206	72,295	4,418	13,010
1995	136,871	44,642	92,229	39,673	77,802	4,969	14,427
1996	153,976	47,635	106,341	42,788	91,851	4,847	14,490
1997	155,546	43,615	111,931	40,562	100,022	3,053	11,909
1998	143,696	37,530	106,166	34,866	94,161	2,664	12,005
1999	133,161	36,565	96,596	33,374	83,412	3,191	13,184
2000 ʳ	156,491	41,250	115,241	37,650	99,942	3,600	15,299
2001	143,750	40,941	102,809	37,679	86,260	3,262	16,549

Source: Bureau of the Census, "Aerospace Industry (Orders, Sales, and Backlog)" (Annually).

U.S. AIRCRAFT PRODUCTION—CIVIL
Calendar Years 1969–2001
(Number of Aircraft)

Year	TOTAL	Domestic Shipments			Export Shipments		
		Trans-ports	Heli-copters	General Aviation	Trans-ports	Heli-copters	General Aviation
1969	13,505	332	282	9,996	182	252	2,461
1970	8,076	127	150	5,246	184	332	2,037
1971	8,158	50	171	5,900	173	298	1,566
1972	10,576	79	319	7,702	148	256	2,072
1973	14,709	143	342	10,482	151	428	3,163
1974	15,326	91	433	9,903	241	395	4,263
1975	15,251	127	528	10,804	188	336	3,268
1976	16,429	64[a]	442	12,232	158	315	3,218
1977	17,913	54	527	13,441	101	321	3,469
1978	18,962	130	536	14,346	111	368	3,471
1979	18,460	176	570	13,177	200	459	3,878
1980	13,634	150	841	8,703	237	525	3,178
1981	10,916	132	619	6,840	255	453	2,617
1982	5,085	111	333	3,326	121	254	940
1983	3,356	133	187	2,172	129	216	519
1984	2,999	102	143	2,013	83	233	425
1985	2,691	126	247	1,545	152	137	484
1986	2,156	171	120	1,031	159	210	464
1987	1,800	187	116	598	170	242	487
1988	1,949	206	103	500	217	280	643
1989	2,448	138	221	225	260	294	1,310
1990	2,268	215	254	335	306	349	809
1991	2,181	204	253	487	385	318	534
1992	1,790	180	112	541	387	212	358
1993	1,630	130	83	631	278	175	333
1994	1,545	87	154	543	222	154	385
1995	1,625	119	82	714	137	210	363
1996	1,662	97	64	732	172	214	383
1997	2,269	122	87	1,140	252	259	409
1998	3,115	184	125	1,794	375	238	399
1999	3,456	279	180	1,972	341	181	503
2000	3,780	217	189	2,391	268	304	411
2001	3,559	273	106	2,172	253	309	446

Source: Aerospace Industries Association, based on company reports; General Aviation Manufacturers Association; and Department of Commerce, International Trade Administration.
   a Prior to 1976, includes the C-130 military transport.

U.S. AIRCRAFT PRODUCTION—MILITARY
Calendar Years 1969–2001
(Number of Aircraft)

Year	TOTAL	U.S. Military Agencies	Exports		
			TOTAL	FMS[a]	Direct[b]
1969	4,290	3,644	646	NA	NA
1970	3,720	3,085	635	NA	NA
1971	2,914	2,232	682	NA	NA
1972	2,530	1,993	537	124	413
1973	1,821	1,243	578	129	449
1974	1,513	799	714	365	349
1975	1,779	844	935	525	410
1976	1,318	625	693	518	175
1977	1,134	454	680	408	272
1978	996	467	529	256	273
1979	837	531	306	203	103
1980	1,047	625	422	194	228
1981	1,062	703	359	215	144
1982	1,159	690	469	68	401
1983	1,053	766	287	70	217
1984	936	561	375	71	304
1985	919	643	276	134	142
1986	1,107	708	399	110	289
1987	1,210	725	485	133	352
1988	1,305	687	618	138	480
1989	1,261	614	647	92	555
1990	1,053	664	387	99	290
1991	911	556	355	94	261
1992	753	422	331	122	209
1993	955[c]	437	518	146	372[c]
1994	764	418	346	69	277
1995	811[d]	354	457	108	349
1996	558	242	316	106	210
1997	488	151	337	181	156
1998	418	149	269	175	94
1999	359	133	226	114	112
2000	333[r]	138	195[r]	42	153[r]
2001	325	176	149	22	127

Source: Aerospace Industries Association, based on USAF, USA, and USN survey responses and Department of Commerce, International Trade Administration.
a Also includes acceptances of NATO AWACS aircraft.
b Military aircraft exported via commercial contracts, directly from manufacturers to foreign governments.
c The number of small (450 kg–2000 kg), new aircraft exported doubled in 1993 to 340 worth $18 million.
d Includes 358 small (450 kg–2000 kg), new aircraft worth $14.7 million.

CIVIL AIRCRAFT SHIPMENTS
Calendar Years 1987–2001

Year	TOTAL	Transport Aircraft[a]	Helicopters	General Aviation
NUMBER OF AIRCRAFT SHIPPED				
1987	1,800	357	358	1,085
1988	1,949	423	383	1,143
1989	2,448	398	515	1,535
1990	2,268	521	603	1,144
1991	2,181	589	571	1,021
1992	1,790	567	324	899
1993	1,630	408	258	964
1994	1,545	309	308	928
1995	1,625	256	292	1,077
1996	1,662	269[a]	278	1,115
1997	2,269	374	346	1,549
1998	3,115	559	363	2,193
1999	3,456	620	361	2,475
2000	3,780	485	493	2,802
2001	3,559	526	415	2,618
VALUE (Millions of Dollars)				
1987	$12,148	$10,507	$277	$1,364
1988	15,855	13,603	334	1,918
1989	17,129	15,074	251	1,804
1990	24,477	22,215	254	2,008
1991	29,035	26,856	211	1,968
1992	30,728	28,750	142	1,836
1993	26,389	24,133	113	2,144
1994	20,666	18,124[E]	185	2,357
1995	18,299	15,263[E]	194	2,842
1996	20,805	17,564[E]	193	3,048
1997	31,753	26,929	231	4,593
1998	41,449	35,663	252	5,534
1999	45,161	38,171	187	6,803
2000	38,637	30,327	270	8,040
2001	42,399	34,155	247	7,997

Source: Aerospace Industries Association, based on company reports and General Aviation Manufacturers' Association.
a U.S.-manufactured fixed-wing aircraft over 33,000 pounds empty weight, including all jet transports plus the four-engine turboprop-powered Lockheed L-100.

SHIPMENTS OF CIVIL TRANSPORT AIRCRAFT[a]
Calendar Years 1997–2001

Company and Model	1997	1998	1999	2000	2001
TOTAL					
Number of Aircraft	374	559	620	485	526
Value (Millions of Dollars)	$26,929	$35,663	$38,171	$30,327	$34,155
Boeing—TOTAL....................	320	505	561	446	475
B-737	135	281	320	278	298
B-747	39	53	47	24	31
B-757	46	50	67	45	45
B-767	41	47	44	44	40
B-777	59	74	83	55	61
Douglas[b]—TOTAL..................	54	54	59	39	51
MD-11	12	12	8	4	2
MD-80	16	8	26	—	—
MD-90	26	34	13	3	—
MD-95 (B-717)	—	—	12	32	49

Source: Aerospace Industries Association, based on company reports.
  a U.S.-manufactured fixed-wing aircraft over 33,000 lbs.
  b Formerly reported as McDonnell Douglas.

SPECIFICATIONS OF U.S. CIVIL JET TRANSPORT AIRCRAFT[a]
On Order or in Production as of 2001

Number of Engines and Crew, and Model Designation[b]	Initial Service	Standard Mixed Class	Operating Empty Weight (000's lbs)	Maximum Takeoff Gross Weight (000's lbs)	Range (Nautical Miles)[c]	Engine Manufacturer[d] and Model
FOUR ENGINES/CREW OF 2						
747-400*	1989	416-524	399	875	7,260	GE CF6-80C2, P&W PW4000, or RR RB211-524
747-400ER*	TBD	416-524	407	910	7,670	GE GE CF6-80C2, P&W PW4000, or RR RB211-524
TWO ENGINES/CREW OF 2						
MD-95 (717)	1999	106	70	121	1,375	BMW-RR BR715
737-600	1998	110-132	80	144	3,050	CFMI CFM56-7B
737-700	1997	126-149	83	155	3,260	CFMI CFM56-7B
737-800	1998	162-189	91	174	2,940	CFMI CFM56-7B
737-900	2001	177-189	95	174	2,745	CFMI CFM56-7B
757-200	1983	200-228	129	255	3,900	RR RB211-535 or P&W PW2000
757-300	1999	243-280	141	273	3,395	RR RB211-535 or P&W PW2000
767-200ER*	1984	181-255	190	395	6,600	P&W PW4000 or GE CF6-80C2
767-300*	1986	218-351	200	412	5,344	P&W PW4000 , GE CF6-80C2, or RR RB211-524
767-300ER*	1997	218-351	203	412	6,105	P&W PW4000, GE CF6-80C2, or RR RB211-524
767-400ER*	2000	245-375	229 227	450	5,645	P&W PW4000 or GE CF6-80C2
777-200*	1995	305-440	310	545	5,210	RR Trent, GE GE90, or P&W PW4000
777-200ER*	1997	301-440	314	656	7,730	RR Trent, GE GE90, or P&W PW4000
777-200LR*	2003	301	344	750	8,865	GE GE90
777-300*	1998	368-550	352	660	5,955	RR Trent, GE GE90, or P&W PW4000
777-300ER*	2003	365	373	750	7,250	GE GE90

Source: Aerospace Industries Association, based on company reports.
    a All jet-powered passenger transport aircraft 33,000 pounds or more empty weight.
    b The Boeing Company manufactures models: 737, 747, 757, 767, & 777 and its Douglas Products Division manufactures MD-95 (renamed the 717).
    c Full passenger load and baggage.
    d P&W = Pratt & Whitney; GE = General Electric; RR = Rolls-Royce; CFMI = General Electric/Snecma; IAE = International Aero Engines; BMW = Bayerische Motoren Werke.
TBD To be determined.
   * Wide-body aircraft.

CIVIL TRANSPORT AIRCRAFT BACKLOG[a]
As of December 31, 1997–2001

Company and Model	1997	1998	1999	2000	2001
TOTAL AIRCRAFT ON ORDER					
Number of Aircraft	1,744	1,786	1,512[b]	1,612[b]	1,357[b]
Value (Millions of Dollars)	$93,788	$86,057	$72,972	$89,780	$75,850
Boeing—TOTAL.....................	1,602	1,595	1,385	1,503	1,313
B-737	909	978	916	1,016	902
B-747	159	102	77	77	62
B-757	133	130	81	79	57
B-767	141	134	122	84	76
B-777	260	251	189	247	216
Douglas[c]—TOTAL.................	142	191	127	109	44
MD-11	14	14	6	2	—
MD-80/90	78	62	3	—	—
MD-95 (B-717)	50	115	118	107	44
TOTAL FOREIGN ORDERS					
Number of Aircraft	790	687	493	477	444
Value[E] (Millions of Dollars)	$51,583	$38,726	$29,939	$38,972	$34,891
Boeing—TOTAL.....................	709	625	458	450	423
B-737	336	344	258	227	229
B-747	122	75	65	64	47
B-757	38	40	15	9	8
B-767	29	21	14	19	21
B-777	184	145	106	131	118
Douglas[c]—TOTAL.................	81	62	35	27	21
MD-11	13	11	6	2	—
MD-80/90	68	36	3	—	—
MD-95 (B-717)	—	15	26	25	21

Source: Aerospace Industries Association, based on company reports.
  a   Unfilled announced orders excluding options for U.S.-manufactured transport aircraft over 33,000 pounds. Includes new transports contracted for lease from the manufacturer.
  b   Includes 84 unidentified orders in 1999, 64 in 2000, and 156 in 2001.
  c   Formerly reported as McDonnell Douglas.

SPECIFICATIONS OF U.S. CIVIL HELICOPTERS
In Production as of 2001

Company	Commercial Model	Number of Places	Useful Load (Lbs.)	Range with Useful Load (N.Miles)	External Cargo Payload (Lbs.)
Brantly International	B-2B	2	620	174[r]	—
Enstrom Helicopter	F-28 Series	3	1,030	241	1,000
	280 Series	3	1,015	260	1,000
	480 Series	5	1,305	375	1,000
Hiller Aircraft	UH-12E3	3	1,341	232	1,000
	UH-12E3T	3	1,460	172	1,000
Kaman	K-1200	1–3[r]	500	267	6,855
MD Helicopters	500 Series	5	1,519	264	2,069
	520 Series	5	1,764	210	2,364
	530F	5	1,509	232	2,159
	600 Series	8	2,000	383	2,720
	900 Series	8	2,875	255	3,466
Robinson Helicopter	R22	2	515	180	—
	R44	4	980	365	—
Schweizer Aircraft	300C	3	950	201	1,050
	300CB	2–3	662	NA	—
	330/333	4	1,300[r]	310[r]	—
Sikorsky Aircraft	S-76C	14	4,813	439	3,300
	S-92	21	10,300	475	10,000

Source: Helicopter Association International, "2002 Helicopter Annual" (Annually).

CIVIL HELICOPTER SHIPMENTS[a]
Calendar Years 1997–2001

Company and Model	1997	1998	1999	2000	2001
CIVIL SHIPMENTS					
Number of Aircraft	346	363	361	493	415
Value (Millions of Dollars)	$231	$252	$187	$270	$247
Brantly—TOTAL	—	2	—	6	2
B-2B	—	2	—	6	2
Enstrom—TOTAL	12	14	8	7	8
F-28/280 series	5	3	5	2	4
480 series	7	11	3	5	4
Hiller—TOTAL	—	—	—	1	2
UH12E	—	—	—	1	2
Kaman—TOTAL	4	2	—	3	6
K-1200	4	2	—	3	6
MD Helicopters[b]—TOTAL	27	37	33	41	28
500 series	9	5	5	11	4
520N series	2	2	5	4	2
530 series	—	5	6	4	—
600 series	15	21	6	8	2
900 series	1	4	11	14	20
Robinson—TOTAL	246	251	278	390	328
R22	132	117	128	126	134
R44	114	134	150	264	194
Schweizer—TOTAL	39	41	35	36	33
300C	15	17	23	13	17
300CB	19	21	11	17	12
330/333	5	3	1	6	4
Sikorsky—TOTAL	18	16	7	9	8
S-70	—	—	—	2	—
S-76	18	16	7	7	8

Source: Aerospace Industries Association, based on company reports.
NOTE: All data exclude production by foreign licensees.
  a Domestic and export helicopter shipments for non-military use. Helicopters in military configuration exported to foreign governments and purchased under commercial contract are reported elsewhere. Please note that shipments from Bell Helicopter's Canadian facilities are excluded as are other foreign-produced helicopters, but reported separately below for information purposes only.
  b Formerly reported as McDonnell Douglas.

	1997	1998	1999	2000	2001
Bell—TOTAL	233	197	146	143	122
206B	35	24	28	14	14
206L/LT	13	16	12	27	10
212	1	1	—	—	—
230	2	2	—	—	—
407	138	104	62	62	47
412	36	35	26	24	22
427	—	—	—	5	15
430	8	15	18	11	14

GENERAL AVIATION AIRCRAFT SHIPMENTS
BY SELECTED MANUFACTURERS
Calendar Years 1997–2001

	1997	1998	1999	2000	2001
NUMBER OF AIRCRAFT SHIPPED	1,549	2,193	2,475	2,802	2,618
Single-Engine, Piston	898	1,434	1,634	1,810	1,581
Multi-Engine, Piston	86	94	114	103	147
Turboprop	223	259	239	315	306
Turbojet	342	406	488	574	584
VALUE[a] OF SHIPMENTS					
(Millions of Dollars)	$4,593	$5,534	$6,803	$8,040	$7,997
Piston	$ 200	$ 330	$ 385	$ 446	$ 471
Turboprop	727	763	658	935	742
Turbojet	3,653	4,441	5,760	6,659	6,784
Number of Aircraft Shipped					
By Selected Manufacturer					
American Champion	46	74	91	96	56
Aviat.....................................	61	85	83	91	57
Bellanca	2	1	1	1	1
Cessna	612	1,072	1,202	1,256	1,202
Cirrus Design	—	—	9	95	183
Classic	6	—	—	—	—
Commander...........................	14	13	13	20	11
Gulfstream	51	61	70	71	71
Lancair	—	—	—	5	27
Learjet	45	61	99	133	109
Maule	54	63	69	57	57
Micco	—	—	—	6	10
Mooney	86	93	97	100	29
Piper	222	295	341	395	441
Raytheon[b]	350	375	400	476	364

Source: General Aviation Manufacturers' Association.
  a Manufacturers' net billing price.
  b Formerly reported as Beech.

DIRECT EXPORT SHIPMENTS OF MILITARY HELICOPTERS[a]
Calendar Years 1997–2001

Manufacturer and Model	1997	1998	1999	2000	2001
TOTAL					
Number of Aircraft	25	50	28	39	10
Value (Millions of Dollars)	$213	$757	$484	$699	$164[b]
Boeing AH-64[c]	—	1	6	1	—
Boeing Vertol CH-47/414/352	1	17	10	7	—
Hiller UH-12E	2	—	—	—	—
Sikorsky S-70C	22	32	12	31	10

Source: Aerospace Industries Association, company reports.
   a Shipments of helicopters in military configuration exported directly from U.S. manufacturers to foreign governments. Military helicopters exported via Foreign Military Sales (FMS) are reported with Dept. of Defense (DoD) aircraft acceptance data elsewhere in this chapter. Some models reported on this page may be shipped in either military or civil configuration; see Civil Helicopter Shipments table for additional data.
   b Estimated by AIA using trade statistics.
   c Formerly reported as McDonnell Douglas.

SPECIFICATIONS OF U.S. MILITARY AIRCRAFT
On Order or in Production as of 2001

Primary Mission, DoD Designation, & Popular Name	Manufacturer	U.S. Military Service	Crew	Empty Weight (000's lbs)	Engines	Performance Typical for Primary Mission	Remarks
ATTACK							
AV-8B Harrier II	Boeing/BAe	USMC	1	15	1xRR F402	Mach 1.0	STOVL Multimission Aircraft
FIGHTERS							
F-15E Eagle	Boeing	USAF	2	37	2xP&W F100	Mach 2.5 class	Dual role fighter/long range interdiction
F-16C/D Fighting Falcon	LM	USAF	1-2	19	1xP&W F100/ 1xGE F110	Mach 2+ class	Multirole fighter
F/A-18C/D Hornet	Boeing/NGC	USN/USMC	1-2	23	2xGE F404	Mach 1.8 class	Multi-mission strike fighter
F/A-18E/F Hornet	Boeing/NGC	USN/USMC	1-2	31	2xGE F414	Mach 1.8 class	Multi-mission strike fighter
F-22A Raptor	LM/Boeing	USAF	1	NA	2xP&W F119	Mach 2+ class	Air superiority with near-precision ground attack
COMMAND/CONTROL AND PATROL							
E-2C Hawkeye 2000	NGC	USN	5	40	2xRR T56	6+ hr. mission duration	AEW command & control; active & passive detection
E-8C Joint STARS	NGC	USAF/Army	21+	171	4xP&W JT3D	11-20+ hr. loiter	Ground surveillance/battle mgmt
RC-12 P/Q	Raytheon	Army	2	9	2xP&W PT6A	4 hr. loiter	Electronic intercept
YAL-1A Airborne Laser	Boeing	USAF	6	TBD	4xGE CF6	TBD	Airborne high-energy chemical laser system
CARGO-TRANSPORT							
C-12R	Raytheon	Army	2	8	2xP&W PT6A	268 mph; 788 n.m.	Utility/transport
C-17A Globemaster III	Boeing	USAF	3	277	4xP&W F117	Mach 0.77; 2,400 n.m.	102 troops or 170,000 lbs.
C-20F/G/H	Gulfstream	All	2	42-43	2xRR Tay	Mach 0.80; 4,200 n.m.	Versions of Gulfstream IV
C-32A	Boeing	USAF	16	132	2xP&W 2040	Mach 0.80; 4,150 n.m.	Executive personnel transport
C-37A	Gulfstream	All	2	48	2xBR 710	Mach 0.80; 6,500 n.m.	Version of Gulfstream V
C-40A	Boeing	USN	3-7	92	2xCFM 56-7	Mach 0.79; 3,000 n.m.	Navy Unique Fleet Essential Aircraft
C-40B	Boeing	USAF	3-7	92	2xCFM 56-7	Mach 0.79; 5,000 n.m.	Special air mission aircraft
C/EC/WC-130J	LM	USAF/ANG	3	97	4xRR AE2100	396 mph; 3,260 mi.	41,000 lbs.
KC-130J	LM	USMC	3	97	4xRR AE2100	12,100 gals.	Tanker
CV/MV-22 Osprey	Bell/Boeing	USMC/USAF	3-4	33	2xRR AE1107C	Max 316 mph; 2,100 n.m.	With internal fuel tanks, engine nacelles tilt for STOL
TRAINING							
T-1A Jayhawk	Raytheon	USAF	2	10	2xP&W JT-15D	Max 538 mph	Tanker/transport trainer
T-6A Texan II	Raytheon	USN/USAF	2	5	1xP&W PT6A-68	Max 368 mph	Primary trainer
T-45C Goshawk	Boeing/BAe	USN	2	11	1xRR F405	Mach 0.85, 675 mph	Next generation trainer
TH-67 Creek	Bell	Army	1	2	1xRR 250	Max 135 mph; 405 mi.	Rotary wing trainer
HELICOPTERS							
AH-1Z	Bell	USMC	2	12	2xGE T700	Max 200 mph; 398 mi.	Attack helicopter
AH-64D Apache	Boeing	Army	2	11	2xGE T700	Max 226 mph; 445 mi.	Attack helicopter
CH-47SD	Boeing	Army	3	25	2xHON T55	Max 178 mph; 750 mi.	Heavy-lift helicopter
CH-53E	Sikorsky	USN	3-8	33-36	3xGE T64	Max 196 mph; 710 mi.	55 passengers, aux. tanks/minesweeping
HH/SH-60 Seahawk	Sikorsky	USN	4-12	14	2xGE T700	Max 184 mph; 500 mi.	Combat search and rescue, SOF
HH/MH-60G Pave Hawk	Sikorsky	USAF	3	12	2xGE T700	Max 184 mph; 1,380 mi.	11 troops; combat; search; rescue
MH-60S	Sikorsky	USN	4	11	2xGE T700	Max 184 mph; 373 mi.	Vertical replenishment
RAH-66 Comanche	Boeing/Sikorsky	Army	2	9	2xLHTEC T801	Max 201 mph; 400 mi.	Armed recon./light attack
UH-1Y	Bell	USMC	2	12	2xGE T700	Max 185 mph; 373 mi.	Utility assault helicopter
UH-60L Black Hawk	Sikorsky	Army	2	11	2xGE T700	Max 184 mph; 373 mi.	Utility assault helicopter

Source: Aerospace Industries Association, based on company reports.
KEY: BAe = BAE Systems; BR = BMW-Rolls Royce; GE = General Electric; HON = Honeywell; LHTEC = Light Helicopter Turbine Engine Co.; LM = Lockheed Martin; NGC = Northrop Grumman; P&W = Pratt & Whitney; RR = Rolls Royce.
TBD To be determined.

MILITARY AIRCRAFT PROGRAM PROCUREMENT[a]
Fiscal Years 2001, 2002, and 2003
(Costs in Millions of Dollars)

Agency and Model	2001		2002 [E]		2003 [E]	
	No.	Cost	No.	Cost	No.	Cost
AIR FORCE						
B-2 Stealth Bomber	—	$ 23.6	—	$ 23.5	—	$ 72.1
C-17 Globemaster III............	12	2,995.0	15	3,762.3	12	3,826.7
C-32B	—	—	1	72.5	—	—
C-130J Hercules..................	3	501.2	6	450.9	2	194.6
Civil Air Patrol Aircraft	50	6.3	27	7.4	27	2.6
E-8C JSTARS	1	286.7	1	317.8	1	279.3
F-15E Eagle	5	661.4	—	241.6	—	232.5
F-16 Falcon.......................	4	411.1	—	232.4	—	265.0
F-22 Raptor	10	2,536.5	13	3,037.3	23	4,621.0
JPATS[b]	58	214.6	46	254.3	35	211.8
Operational Support Aircraft	1	69.2	—	—	—	—
Unmanned Aerial Vehicles[c] ...	11	88.4	27	451.4	37	425.6
ARMY						
AH-64 Apache Mods	52	$ 755.2	60	$ 910.8	74	$ 895.5
C-XX Medium Range Aircraft	1	7.5	1	45.0	—	—
CH-47 Mods	—	182.3	—	269.1	—	403.9
TH-67 Creek	17	23.8	15	25.0	—	—
UH-60 Black Hawk	18	211.3	12	199.5	12	180.2
NAVY						
AV-8B Harrier.....................	12	$ 259.8	—	$ —	—	$ 6.0
C-37..............................	1	49.7	—	—	—	—
C-40A[b]............................	2	122.0	—	—	—	—
E-2C Hawkeye	5	312.4	5	275.2	5	295.5
EA-6B Prowler	—	184.4	—	149.7	—	223.5
F/A-18E/F Hornet	39	2,837.8	48	3,118.3	44	3,159.5
KC-130J	3	227.3	2	154.8	4	334.0
MH-60R	—	53.7	—	9.9	—	116.2
MH-60S	15	283.8	13	254.0	15	372.2
T-45 Goshawk	14	302.3	6	183.4	8	221.4
UC-35	1	7.5	1	7.4	—	—
V-22 Osprey[b]	9	1,170.5	11	942.3	11	1,497.2

Source: Department of Defense Budget, "Program Acquisition Costs by Weapon System" (Annually) and "Procurement Programs (P-1)" (Annually).

NOTE: See Research and Development Chapter for aircraft program RDT&E authorization data.

    a Total Obligational Authority for procurement, excluding initial spares and mods (except where noted).

    b Air Force and Navy funding.

    c Air Force and Army funding.

DEPARTMENT OF DEFENSE OUTLAYS FOR AIRCRAFT PROCUREMENT
BY AGENCY
Fiscal Years 1968–2003
(Millions of Dollars)

Year	TOTAL	Air Force	Army	Navy
1968	$ 9,462	$ 5,079	$1,139	$ 3,244
1969	9,177	5,230	1,126	2,821
1970	7,948	4,623	2,488	837
1971	6,631	3,960	546	2,125
1972	5,927	3,191	389	2,347
1973	5,066	2,396	113	2,557
1974	5,006	2,078	122	2,806
1975	5,484	2,211	136	3,137
1976	6,520	3,323	136	3,061
Tr.Qtr.	1,557	859	26	672
1977	6,608	3,586	301	2,721
1978	6,971	3,989	380	2,602
1979	8,836	5,138	558	3,140
1980	11,124	6,647	787	3,689
1981	13,193	7,941	855	4,397
1982	16,793	9,624	1,297	5,872
1983	21,013	11,799	1,724	7,490
1984	23,196	12,992	2,165	8,040
1985	26,586	15,619	2,705	8,263
1986	30,828	18,919	2,987	8,922
1987	32,956	20,036	3,306	9,614
1988	28,246	15,961	2,878	9,407
1989	27,569	14,662	2,834	10,073
1990	26,142	14,303	2,808	9,031
1991	25,689	13,794	2,840	9,055
1992	23,581	13,154	2,520	7,907
1993	20,359	11,438	1,675	7,246
1994	18,840	10,303	1,711	6,826
1995	16,125	8,891	1,549	5,685
1996	14,331	7,862	1,435	5,034
1997	14,663	7,799	1,542	5,322
1998	15,473	8,236	1,392	5,845
1999	16,484	8,928	1,532	6,024
2000	17,991	8,979	1,268	7,744
2001	17,979	8,217	1,358	8,404
2002 [E]	19,496	9,988	1,699	7,809
2003 [E]	20,023	10,194	1,814	8,015

Source: Office of Management and Budget, "Budget of the United States Government" (Annually).

MILITARY AIRCRAFT ACCEPTED BY U.S. MILITARY AGENCIES
Calendar Years 1987–2001

Year	TOTAL	Bomber/ Patrol/ Command/ Control	Fighter/ Attack	Trans- port/ Tanker	Trainer	Heli- copter	Other
NUMBER OF AIRCRAFT							
1987	858	74	483	36	—	265	—
1988	842	55	509	31	—	247	—
1989	706	24	408	21	—	253	—
1990	763	24	454	25	—	260	—
1991	650	17	395	23	—	215	—
1992	544	10	312	30	37	155	—
1993	583	11	293	25	56	198	—
1994	487	6	167	40	114	157	3
1995	462	4	133	32	102	176	15
1996	348	4	116	28	54	146	—
1997	332	4	202	19	26	81	—
1998	324	10	188	30	33	63	—
1999	247	6	153	45	12	31	—
2000 ʳ	180	2	82	30	33	33	—
2001	198	4	74	27	52	41	—
FLYAWAY VALUE (Millions of Dollars)							
1987	$21,459	$8,569	$8,900	$2,218	$ —	$1,772	$ —
1988	16,031	2,911	8,953	2,314	—	1,853	—
1989	11,968	1,423	7,735	743	—	2,067	—
1990	13,036	1,499	8,731	605	—	2,201	—
1991	11,754	1,023	8,517	437	—	1,777	—
1992	11,482	613	7,673	1,346	267	1,583	—
1993	12,101	1,530	6,400	1,553	484	2,134	—
1994	13,000	3,861	3,661	3,298	477	1,686	17
1995	12,369	3,585	3,547	2,759	460	1,922	98
1996	11,383	3,596	3,524	2,350	337	1,576	—
1997	10,945	1,921	5,653	2,336	270	766	—
1998	15,099	4,831	6,240	2,890	319	835	—
1999	10,731	415	6,164	3,588	219	345	—
2000 ʳ	8,366	140	3,810	3,651	356	409	—
2001	7,506	287	3,634	2,735	376	474	—

Source: Aerospace Industries Association, based on USAF, USA, and USN survey responses.
NOTE: Data represent new U.S.-manufactured aircraft, excluding gliders and targets. Values include spares, spare parts, and support equipment that are procured with the aircraft. Includes aircraft accepted for shipment to foreign governments for military assistance programs and foreign military sales.

MILITARY AIRCRAFT ACCEPTANCES BY UNITED STATES AIR FORCE[a]
Calendar Years 2000–2001
(Costs in Millions of Dollars)

Type and Model	Number of Aircraft		Flyaway Cost[b]		Weapon System Cost[c]	
	2000	2001	2000	2001	2000	2001
TOTAL................................	47[r]	60	$3,408[r]	$2,565	$3,799[r]	$3,202
Fighter/Attack—TOTAL	10	5	$ 355[r]	$ 123	$ 365[r]	$ 123
F-15	5	—	232	—	236	—
F-16	5	5	123[r]	123	129[r]	123
Transports/Tankers—TOTAL	20[r]	18	2,972[r]	2,330	3,325[r]	2,920
C-17	13	10	2,660	1,930	2,943	2,250
C-37	2	2[d]	75	88	79	92
C-130 variants..................	5[r]	6	237[r]	312	304[r]	578
Trainers—TOTAL	17	37	81	112	110	159
T-6................................	17	37	81	112	110	159

Source: Aerospace Industries Association, based on USAF survey responses.
   a   Air Force acceptances for own use; excludes FMS/MAP shipments.
   b   Flyaway Cost includes airframe, engines, electronics, communications, armament, other installed equipment, and non-recurring costs associated with the manufacture of aircraft.
   c   Weapon System Cost includes flyaway costs, peculiar ground equipment, training equipment, and technical data.
   d   Under Lease.

MILITARY AIRCRAFT ACCEPTANCES BY UNITED STATES ARMY[a]
Calendar Years 2000–2001
(Costs in Millions of Dollars)

Type and Model	Number of Aircraft		Flyaway Cost[b]		Weapon System Cost[c]	
	2000	2001	2000	2001	2000	2001
TOTAL..............................	14	22	$116[r]	$193	$132[r]	$230
Transports/Tankers—TOTAL	2	—	$ 12	$ —	$ 12	$ —
UC-35	2	—	12	—	12	—
Helicopters—TOTAL............	12	22	104[r]	193	120[r]	230
UH-60L	12	22	104[r]	193	120[r]	230

Source: Aerospace Industries Association, based on USA survey responses.
a Army acceptances for own use; excludes FMS/MAP shipments.
b Flyaway Cost includes airframes, engines, electronics, communications, armament and other installed equipment.
c Weapon System Cost includes flyaway cost, initial spares, ground equipment, training equipment and other support items.

MILITARY AIRCRAFT ACCEPTANCES BY UNITED STATES NAVY[a]
Calendar Years 2000–2001
(Costs in Millions of Dollars)

Type and Model	Number of Aircraft		Flyaway Cost[b]		Weapon System Cost[c]	
	2000	2001	2000	2001	2000	2001
TOTAL.............................	77[r]	94	$3,901[r]	$4,162	$4,344[r]	$5,068
Patrol—TOTAL	2	4	$ 140[r]	$ 287	$ 159[r]	$ 312
E-2.................................	2	4	140[r]	287	159[r]	312
Fighter/Attack—TOTAL	41[r]	50	2,663[r]	2,966	2,894[r]	3,709
AV-8B	11[r]	14	237[r]	286	301[r]	434
F/A-18	30	36	2,427	2,680	2,593	3,275
Transports/Tankers—TOTAL	8	9	667	405	769	446
C-40	—	4	—	199	—	207
KC-130.........................	1	3	54	195	68	227
UC-35	—	2	—	10	—	2
V-22	7	—	613	—	701	—
Trainers—TOTAL	16	15	276	264	299	307
T-45A	16	15	276	264	299	307
Helicopters—TOTAL...........	10[r]	16	155[r]	240	222[r]	294
MH-60 S	10[r]	16	155[r]	240	222[r]	294

Source: Aerospace Industries Association, based on USN survey responses.
a Navy acceptances for own use; excludes FMS shipments.
b Flyaway Cost includes airframe, engines, electronics, communications, armament, other installed equipment, non-recurring costs, and ancillary equipment.
c Weapons System Cost (Investment Cost) includes flyaway cost, initial spares, ground equipment, training equipment, and other support items.

MILITARY AIRCRAFT ACCEPTANCES FOR REIMBURSABLE PROGRAMS[a]
Calendar Years 2000–2001
(Costs in Millions of Dollars)

Accepting Agency, Type, and Model	Number of Aircraft		Flyaway Cost[b]	
	2000	2001	2000	2001
TOTAL	42	22	$942 [r]	$587
AIR FORCE—TOTAL...............	20	19	$396	$546
Fighter/Attack—TOTAL	20	19	396	546
F-16	20	19	396	546
NAVY—TOTAL	22	3	$547	$ 41
Fighter/Attack—TOTAL	11	—	396	—
F/A-18	11	—	396 [E]	—
Helicopters—TOTAL..............	11	3	151	41
AH-1	11	3	151	41

Source: Aerospace Industries Association, based on USAF, USA, and USN survey responses.

   a Foreign government aircraft purchases through the Department of Defense Foreign Military Sales program.

   b Flyaway Cost includes airframes, engines, electronics, communications, armament, other installed equipment, and non-recurring costs associated with the manufacture of the aircraft.

MISSILE PROGRAMS

Sales of missile systems and parts reached an eight-year high in 2001, coming in at $6.2 billion. The Census Bureau also reported that new orders fell nearly 50% after more than doubling in the previous year. The year-end 2001 backlog of unfilled orders stood at $8.4 billion.

DoD outlays for missile procurement rose for the second straight year in FY 2001 following a nine-year slide. Despite the 29% increase to $5.4 billion, procurement spending will remain much lower than in the late 1980s and early 1990s. Ballistic Missile Defense continues to dominate missile RDT&E. Funding totaled $4.2 billion in FY 2001 and is scheduled to increase to $7.0 billion in 2002.

Each service funded procurement in excess of $200 million for a few programs. For instance, the Air Force's largest program was the Joint Direct Attack Munition (JDAM) with $273 million in combined USAF/USN funding. The joint procurement totaled 10,976 units, which use a low-cost Global Positioning System unit and some additional fins to turn a simple gravity bomb into a precision-guided munition. The Navy's Trident II, with $437 million of funding for 12 submarine-launched ballistic missiles, constituted its largest missile procurement program; next was 104 JSOWs (Joint Standoff Weapon), with $216 million in Navy/USAF funding. The Army had four large programs: Javelin, a joint Army/USMC missile, with $348 million in combined funding; Advanced Tactical Missile System (ATACMS), $311 million; Hellfire, a joint Army/Navy missile, $303 million; and the Multiple Launch Rocket System, $203 million. Finally, the BMDO funded the purchase of 40 Patriots with joint BMDO/Army money totaling $385 million.

Procurement of several smaller programs continued. The Air Force had three other missile procurements underway in FY 2001: 233 Advanced Medium Range Air-to-Air Missiles (AMRAAM) costing $133 million; 300 Sensor-Fused Weapons (SFW), $112 million; and 5,918 Wind-Corrected Munitions Dispensers (WCMD), $100 million. The Navy had four other programs in active procurement including: 86 Standard missiles costing $172 million; 307 Short-Range Antitank Weapons (SRAW) for $43 million; 29 Evolved SeaSparrow Missiles (ESSM) for $39 million; and 30 Sea-Launched Attack Missile-Extended Range (SLAM-ER), $24 million. The Army bought six Avengers worth $30 million.

In FY 2002, five programs begin or return to unit procurement: AIM-9X, the joint Air Force/Navy advanced air-to-air missile with 243 costing $63 million; 76 Air Force JASSMs for $45 million; 346 Army Stingers for $34 million; 90 Navy RAMs for $43 million; and 32 Navy Tomahawks at $74 million.

ORDERS, SALES, AND BACKLOG OF MISSILE SYSTEMS AND PARTS[a]
Calendar Years 1987–2001
(Millions of Dollars)

Year	SALES—Current Dollars	SALES—Constant Dollars[b]
1987	$ 9,671	$ 9,671
1988	9,485	9,345
1989	9,283	8,799
1990	9,102	8,381
1991	8,989	8,069
1992	9,032	7,726
1993	7,713	6,417
1994	5,294	4,315
1995	4,688	3,765
1996	4,792	3,782
1997	4,024	3,149
1998	4,356	3,390
1999	4,521	3,494
2000[r]	5,567	4,176
2001	6,231	4,558

Year	NET NEW ORDERS	BACKLOG AS OF DECEMBER 31
1987	$11,482	$14,302
1988	9,437	14,255
1989	8,998	14,005
1990	7,917	12,956
1991	8,072	12,571
1992	9,234	11,814
1993	4,775	9,305
1994	2,785	5,823
1995	3,164	4,833
1996	8,672	6,563
1997	4,239	5,828
1998	4,884	6,539
1999	3,753	5,342
2000[r]	9,738	9,389
2001	5,211	8,370

Source: Bureau of the Census, "Aerospace Industry (Orders, Sales, and Backlog)" (Annually).
   a   Excludes engines and propulsion units where separable.
   b   Based on AIA's aerospace composite price deflator, 1987=100.

DEPARTMENT OF DEFENSE OUTLAYS FOR MISSILE PROCUREMENT BY AGENCY

Fiscal Years 1968–2003
(Millions of Dollars)

Year	TOTAL	Air Force	Army	Navy
1968	$ 2,219	$1,388	$ 395	$ 436
1969	2,509	1,382	593	534
1970	2,912	1,467	743	702
1971	3,140	1,497	852	791
1972	3,009	1,334	844	831
1973	3,023	1,454	941	628
1974	2,981	1,537	903	541
1975	2,889	1,602	672	615
1976	2,296	1,549	163	584
Tr.Qtr.	402	347	(93)	148
1977	2,781	1,501	374	905
1978	3,096[a]	1,376	418	1,302[a]
1979	3,786	1,537	547	1,702
1980	4,434	1,810	651	1,973
1981	5,809	2,366	1,146	2,297
1982	6,782	3,069	1,269	2,444
1983	7,795	3,383	1,600	2,812
1984	9,527	4,640	2,079	2,809
1985	10,749	5,409	2,399	2,941
1986	11,731	6,473	2,478	2,780
1987	11,473	6,002	2,314	3,157
1988	11,676	6,046	2,239	3,392
1989	14,503	7,349	2,709	4,445
1990	14,851	7,951	2,453	4,446
1991	14,400	6,906	2,540	4,954
1992	13,504	6,409	2,401	4,694
1993	11,404	5,424	2,187	3,794
1994	8,934	4,312	1,384	3,238
1995	7,513	3,845	974	2,694
1996	6,199	3,235	919	2,045
1997	5,225	2,743	936	1,546
1998	4,907	2,543	964	1,400
1999	4,080	2,299	783	998
2000	4,231	2,243	926	1,062
2001	5,441	2,982	1,248	1,211
2002 [E]	4,923	2,560	1,283	1,080
2003 [E]	5,389	2,987	1,258	1,144

Source: Office of Management and Budget, "The Budget of the United States Government" (Annually).
   a   Beginning 1978, DoD combined Navy Missile Procurement with torpedoes and other related products into Navy Weapons Procurement. Missiles comprise approximately 80 percent of the value of this category.

MAJOR MISSILE PROGRAMS
IN RESEARCH, DEVELOPMENT, OR PRODUCTION
As of 2001

Program	Agency	Status	Systems Contractor	Propulsion Manufacturer	Guidance Manufacturer
AIR-TO-AIR					
AMRAAM-120B/C	USAF/USN	P	Raytheon	ATK	Raytheon/NGC
Sidewinder-9M	USN/USAF	P	NASC	ATK	Raytheon
Sidewinder-9X	USN/USAF	P	Raytheon	ATK	Raytheon
AIR-TO-SURFACE					
AGM-142	USAF	P	LM/Rafael	Rafael	NGC/BAE
GATS/GAM	USAF	P	NGC	—	Honeywell
HARM-88A/B	USN/USAF	P	Raytheon	ATK	Raytheon
*Harpoon-84A/C/ D/G	USN	P	Boeing	TCM/ATK	Ray/Kearfott/ IBM/LSI
*Harpoon-84L	USN	P	Boeing	TCM/ATK	HI/IBM/LSI/Ray/ Kearfott
JASSM	USN/USAF	D	LM	TCM	HI/NGC
JDAM	USAF/USN	D	Boeing	—	HI/Boeing
JSOW-154	USN/USAF	D	Raytheon	—	Kearfott
Maverick-65D/G/H/K	USAF	P	Raytheon	ATK	Raytheon
Maverick-65F	USN	P	Raytheon	ATK	Raytheon
Maverick-65J	USN/USMC	D	Raytheon	ATK	Raytheon
Paveway-Enhanced	USN/USAF	P	Raytheon	—	Raytheon
SLAM-ER-84H/K	USN	P	Boeing	TCM	Boeing/Ray/HI
WCMD	USAF	P	LM	—	LM/BAE
ANTI-SUBMARINE					
VLA-44A	USN	P	LM	ATK	LM
BATTLEFIELD SUPPORT AND ANTIARMOR					
ATACMS	Army	P	LM	ARC	Honeywell
BAT	Army	P	NGC	—	NGC/Ray/BAE
GMLRS	Army	D	LM	ARC	NGC/HI
HELLFIRE II-114K	Army/USMC	P	LM	ATK	LM
Longbow HELLFIRE 114L	Army	P	LM/NGC	ATK	LM/NGC/BAE
HELLFIRE-114M	USN/USMC	P	LM	ATK	LM
LOSAT	Army	D	LM	ATK/ARC	Ray/HI
Javelin	Army/USMC	P	Ray/LM	ARC	LM/Ray/BAE
MLRS-26	Army	P	LM	ARC	—
PGMM	Army	D	LM	—	HI/Draper
Predator	USMC	P	LM	ATK	LM
TOW2A-71E	Army	P	Raytheon	ATK	Raytheon
TOW2B-71F	Army	P	Raytheon	ATK	Raytheon

* Also Surface-to-Surface

(Continued on next page)

MAJOR MISSILE PROGRAMS
IN RESEARCH, DEVELOPMENT, OR PRODUCTION
As of 2001, continued

Program	Agency	Status	Systems Contractor	Propulsion Manufacturer	Guidance Manufacturer
SURFACE-TO-AIR					
GMD	Army	R,D	Boeing	ATK/UTC	Boeing
PAC-3	Army	P	LM	ARC	LM/HI/ Boeing
RAM-116A	USN	P	Raytheon	ARC	Ray/RAMSYS
RAM-116B	USN	P	Raytheon	ARC	Ray/RAMSYS
SeaSparrow-7M	USN	P	Raytheon	ATK	Raytheon
SeaSparrow- Evolved	USN	D	Raytheon	ATK/Raufoss	Raytheon/HI
Standard 2 MR	USN	P	Raytheon	ARC	Raytheon
Standard 2-IV	USN	P	Raytheon	ARC/UTC	Raytheon
Standard 3	USN	D	Raytheon	ARC/UTC/ATK	Raytheon
Stinger-92D/E	All	P	Raytheon	ARC	Raytheon
THAAD	Army	D	LM	UTC/Boeing	LM
SURFACE-TO-SURFACE					
*Harpoon-84A/C/D	USN	P	Boeing	TCM/ATK	Ray/IBM/LSI/ HI/Kearfott
*Harpoon-84L	USN	D	Boeing	TCM/ATK	HI/IBM/LSI/Ray/ Kearfott
Minuteman III	USAF	P	TRW	ATK/UTC/ARL	Boeing
Tomahawk Tactical	USN	D	Raytheon	WI/ARC	Ray/HI
Trident 2 (D-5)	USN	P	LM	ATK/ARC	LM/Draper/ Ray/Boeing/ Kearfott

Source: Aerospace Industries Association, based on company reports.
Status: R-Research; D-Development; P-Production.
* Also Air-to-Surface

Abb:					
ARC	— Atlantic Research	LSI	— Lear Siegler	Ray	— Raytheon
ATK	— Alliant Techsystems	LM	— Lockheed Martin	TCM	— Teledyne Continental Motors
BAE	— BAE Systems	NASC	— Naval Air Systems Command	UTC	— United Technologies
HI	— Honeywell	NGC	— Northrop Grumman	WI	— Williams International

MISSILE PROGRAM PROCUREMENT[a]
Fiscal Years 2001, 2002, and 2003
(Costs in Millions of Dollars)

Agency and Model	2001		2002 [E]		2003[E]	
	No.	Cost	No.	Cost	No.	Cost
AIR FORCE						
AIM-9X[b]	—	$ —	243	$ 62.9	581	$110.3
AMRAAM[b]	233	133.3	247	140.7	261	139.5
JASSM	—	0.2	76	44.7	100	54.2
JDAM[b]	10,976	272.7	22,800	668.4	35,000	764.9
SFW	300	112.0	263	108.8	298	106.0
WCMD	5,918	100.3	6,917	111.4	4,959	71.2
ARMY						
ATACMS	134	$310.5	18	$ 85.7	—	$ 58.8
Avenger	6	29.5	—	11.5	—	—
Hellfire[c]	2,200	302.5	2,200	240.1	1,797	184.4
HIMARS[c]	—	—	—	—	36	136.3
Javelin[d]	3,081	348.1	4,139	411.9	1,725	250.6
LOSAT	—	—	—	9.4	144	17.9
MLRS/GMLRS	—	202.6	—	137.1	5,754	186.7
Stinger[d]	—	14.7	346	34.2	160	32.5
BMDO						
Patriot[f]	40	$385.0	72	$756.7	72	$663.7
TMD BMC3	—	3.9	—	—	—	—
NAVY						
ESSM	29	$ 39.4	26	$ 41.7	146	$129.6
JSOW[b]	104	216.1	35	29.4	476	195.2
RAM	—	22.7	90	42.7	90	58.4
SLAM-ER	30	23.8	30	26.0	120	83.8
SRAW	307	43.0	—	—	445	36.5
Standard	86	171.9	96	156.2	93	156.4
Tomahawk	—	—	32	74.0	106	145.8
Trident II	12	436.5	12	538.3	12	585.9

Source: Department of Defense, "Program Acquisition Costs by Weapon System" (Annually) and "Procurement Programs (P-1)" (Annually).
a Total Obligational Authority excluding initial spares and RDT&E.
b Navy and Air Force funding.
c Army and Navy funding.
d Army and Marine Corps funding.
f Army and BMDO funding.

MISSILE PROGRAMS RESEARCH, DEVELOPMENT, TEST, AND EVALUATION[a] BY AGENCY AND MODEL

Fiscal Years 2001, 2002, and 2003
(Millions of Dollars)

Agency and Model	2001	2002[E]	2003[E]
AIR FORCE			
ACM/ALCM..	$ 9.9	$ 9.2	$ 29.5
AMRAAM[b]	61.7	67.8	45.1
ICBM	62.4	127.8	196.3
*JASSM[b]	112.5	81.1	57.0
JDAM[b]	38.7	83.4	65.3
TSSAM[b]	112.6	81.1	57.0
ARMY			
AAWS-M	$ 0.5	$ 2.8	$ 0.5
BAT (ATACMS)	97.9	122.9	190.3
LOSAT	25.4	21.4	14.5
MEADS	—	0.1	117.7
MLRS	63.0	99.5	57.8
Patriot	12.4	13.8	194.5
MDA			
BMD	$4,208.4	$6,969.4	$6,690.7
NAVY			
AIM-9X Sidewinder[b]	$ 45.4	$ 21.9	$ 4.8
HARM	38.5	28.1	60.8
JSOW[b]	28.0	26.6	16.7
SRAW	11.3	10.7	8.1
Standard	0.5	13.9	16.3
Tomahawk	92.5	75.3	94.3
Trident II	50.1	45.4	40.3

Source: Department of Defense Budget, "Program Acquisition Costs by Weapon System" (Annually) and "RDT&E Programs (R-1)" (Annually).
a Total Obligational Authority.
b Navy and Air Force funding.
* Programs in R&D only.

Missile Program Acronyms:

AAWS-M	—Advanced Anti-tank Weapon System-Medium	ACM/ALCM	—Advanced & Air Launched Cruise Missile
AMRAAM	—Advanced Medium Range Air-to-Air Missile	ATACMS	—Army TACtical Missile System
BAT	—Brilliant Anti-Tank submunition	BMD	—Ballistic Missile Defense
HARM	—High-speed Anti-Radiation Missile	ICBM	—Inter-Continental Ballistic Missile
JASSM	—Joint Air-to-Surface Stand-Off Missile	JDAM	—Joint Direct Attack Munition
JSOW	—Joint Stand-Off Weapon	LOSAT	—Line of Sight Anti-Tank
MEADS	—Medium Extended Air Defense System	MLRS	—Multiple Launch Rocket System
SFW	—Sensor-Fused Weapon	SRAW	—Short Range Attack Weapon
TSSAM	—Tri-Service Standoff Attack Missile	WCMD	—Wind-Corrected Munitions Dispenser

BALLISTIC MISSILE DEFENSE FUNDING
Fiscal Years 1999–2003
(Millions of Dollars)

Category and Title	Program Element	1999	2000	2001	2002	2003 [E]
TOTAL		$3,977	$4,711	$5,888	$8,351	$6,714
BMDO Programs—TOTAL		$3,547	$3,669	$4,755	$8,351	$6,714
Procurement—TOTAL		$ 369	$ 416	$ 444	$ 684	$ —
TMD BMC3	0208864C	23	—	4	—	—
PAC-3	0208865C	303	361	365	677	—
Navy Area	0208867C	43	55	—	7	—
National Missile Defense	0208871C	—	—	75	—	—
Military Construction—TOTAL ...		10	17	104	8	23
National Missile Defense	0603871C	10	17	104	—	—
BMD System	0603880C	—	—	—	8	—
Terminal	0603881C	—	—	—	1	23
RDT&E—TOTAL		3,168	3,236	4,207	7,659	6,691
TMD—TOTAL.......................		1,559	1,830	1,653	—	—
THAAD System	0603861C	527	528	—	—	—
Navy Theater Wide	0603868C	310	420	463	—	—
MEADS	0603869C	24	49	53	—	—
Joint Theater Missile Defense...	0603872C	177	201	—	—	—
THAAD System	0604861C	—	—	550	—	—
PAC-3	0604865C	177	181	81	—	—
Navy Area..........................	0604867C	246	310	274	—	—
Family of Systems E&I............	0603873C	97	142	231	—	—

(Continued on next page)

BALLISTIC MISSILE DEFENSE FUNDING
Fiscal Years 1999–2003, continued
(Millions of Dollars)

Category and Title	Program Element	1999	2000	2001	2002	2003 [E]
RDT&E (continued)						
NMD......................................	0603871C	$950	$ 852	$1,875	$ —	$ —
Support Technology/Applied Research	0602173C	101	84	40	—	—
Support Technology/Advanced Development	0603173C	283	214	122	—	—
Support Technology/Space Based Laser	0603174C	—	—	75	—	—
BMD Technology	0603175C	—	—	—	113	122
International Cooperative Programs	0603875C	63	37	125	—	—
BMD Technical Operations	0603874C	190	204	291	—	—
Threat and Countermeasures......	0603876C	22	16	23	—	—
BMD System	0603880C	—	—	—	780	1,066
Terminal Defense Segment	0603881C	—	—	—	1,610	170
Midcourse Defense Segment......	0603882C	—	—	—	3,941	3,193
Boost Defense Segment	0603883C	—	—	—	685	797
Sensors Segment	0603884C	—	—	—	496	373
Ground-Based Terminal...........	0604861C	—	—	—	—	935
Pentagon Reservation Maintenance Fund	0901585C	—	—	5	7	7
Headquarters Management	0901598C	—	—	—	28	28
BMD-Related Programs—TOTAL		$430	$1,042	$1,133	$ —	$ —
Air Force SBIR (SMTS)	0603441F	$160	$ 229	$ —	$ —	$ —
Air Force SBIRS-High	0604441F	—	421	569	—	—
Air Force SBIRS-Low	0604442F	—	—	241	—	—
Air Force ABL	0603319F	235	309	234	—	—
Air Force SBL	0603876F	35	74	73	—	—
Army THEL	0602307A	—	10	16	—	—

Source: Ballistic Missile Defense Organization.

Space Programs

Sales of space vehicle systems fell for the fourth straight year in 2001 to $7.8 billion, according to figures compiled by the U.S. Census Bureau. That small ($372 million) overall decline, however, masked a 21% drop in the non-military sector, which saw sales fall $919 million to $3.5 billion, largely offsetting a $547 million gain in military sales.

Net new orders fell 29% to $5.1 billion, as 56% stronger military orders of $3.6 billion could not offset the sharp drop-off in civil orders to $1.5 billion from $4.9 billion only a year earlier. The resulting backlog of unfilled orders declined $2.9 billion to $18 billion with a 45/55 split between military and civil work.

A variety of unclassified military space programs received funding in FY 2001. The Air Force led the majority of major DoD space programs, including: EELV for $664 million; SBIRS-High, $550 million; Titan Launch Vehicles, $415 million; the "MilSatCom, EHF" Program, $254 million; NAVSTAR GPS, $401 million; SBIRS-Low, $234 million; and Milstar, $225 million. The Army's DSCS received $109 million of Army/Air Force funding and the Navy's Sat Comm Systems and FleetSatComm programs received $234 million and $95 million, respectively. Of the above-mentioned programs, only "MilSat Com, EHF" is scheduled to receive increased funding in FY 2002.

Federal outlays on space programs totaled $27 billion in FY 2001, with DoD accounting for $13.0 billion and NASA $13.2 billion. The Commerce Department's space outlays rose minimally in 2001 to $519 million while the remainder of the federal government's space spending was relatively flat.

Building the International Space Station and operating the Space Shuttle dominated NASA's FY 2001 budget, accounting for 37% of NASA's $14.3 billion in budget authority for the year. The Station accounted for $2.1 billion in budget authority while operating the

Shuttle cost $3.1 billion, according to NASA budget summaries. Space science funding totaled $2.6 billion and aerospace technology received $2.2 billion. NASA's FY 2002 budget should total $14.9 billion and all the above categories are scheduled to receive increased funding except the International Space Station, whose funding was cut 19% to $1.7 billion.

While the pace of launching has decreased from 2000 for all countries, Commonwealth of Independent States (formerly U.S.S.R.) has managed substantial market share gains over the last four years—rising from 25% to 47%.

Orders for commercial geosynchronous (GEO) satellites fell in 2001, according to figures supplied by the Futron Corporation. The world's manufacturers received 25 orders worth $2.4 billion in 2001—down from 31 worth $3.3 billion in 2000. The unfilled order backlog was estimated at 84 commercial GEO satellites worth $9.2 billion.

ORDERS, SALES, AND BACKLOG OF SPACE VEHICLE SYSTEMS[a]
Calendar Years 1987–2001
(Millions of Dollars)

Year	SALES—Current Dollars			SALES—Constant Dollars[b]		
	TOTAL	Military	Non-Military	TOTAL	Military	Non-Military
1987	$ 8,051	$5,248	$ 2,803	$ 8,051	$ 5,248	$ 2,803
1988	8,622	6,190	2,432	8,495	6,099	2,396
1989	9,758	6,457	3,301	9,249	6,120	3,129
1990	9,691	6,556	3,135	8,924	6,037	2,887
1991	10,515	6,770	3,745	9,439	6,077	3,362
1992	9,266	5,887	3,379	7,926	5,036	2,891
1993	7,317	4,175	3,142	6,087	3,473	2,614
1994	10,594	5,707	4,887	8,634	4,651	3,983
1995	11,314	4,782	6,532	9,088	3,841	5,247
1996	11,698	5,613	6,085	9,233	4,430	4,803
1997	14,643	4,919	9,724	11,458	3,849	7,609
1998	9,491	4,227	5,264	7,396	3,289	4,096
1999	9,022	5,107	3,915	6,972	3,947	3,026
2000[r]	8,164	3,723	4,441	6,125	2,793	3,332
2001	7,792	4,270	3,522	5,700	3,124	2,576

Year	NET NEW ORDERS			BACKLOG AS OF DECEMBER 31		
	TOTAL	Military	Non-Military	TOTAL	Military	Non-Military
1987	$11,455	$9,000	$ 2,455	$12,393	$ 9,460	$ 2,933
1988	7,296	4,561	2,735	10,838	7,880	2,958
1989	11,709	8,107	3,602	13,356	9,192	4,164
1990	9,598	6,256	3,342	12,462	8,130	4,332
1991	11,222	5,468	5,754	11,664	6,221	5,443
1992	10,491	6,773	3,718	12,809	7,622	5,187
1993	8,436	5,106	3,330	13,663	7,384	6,279
1994	9,041	4,896	4,145	12,888	6,732	6,156
1995	13,212	4,679	8,533	15,650	5,872	9,778
1996	16,527	8,888	7,639	23,004	9,125	13,879
1997	15,078	4,584	10,494	23,189	8,848	14,341
1998	12,420	4,563	7,857	20,372	7,970	12,402
1999	11,175	7,912	3,263	22,356	10,666	11,690
2000[r]	7,205	2,310	4,895	21,395	8,942	12,453
2001	5,112	3,605	1,507	18,479	8,277	10,202

Source: Bureau of the Census, "Aerospace Industry (Orders, Sales, and Backlog)" (Annually).
   a Excludes engines and propulsion units where separable.
   b Based on AIA's aerospace composite price deflator, 1987=100.

ORDERS, SALES, AND BACKLOG OF ENGINES AND PROPULSION UNITS FOR MISSILES AND SPACE VEHICLES

Calendar Years 1987–2001
(Millions of Dollars)

Year	SALES—Current Dollars			SALES—Constant Dollars[a]		
	TOTAL	Military	Non-Military	TOTAL	Military	Non-Military
1987	$2,993	$1,563	$1,430	$2,993	$1,563	$1,430
1988	3,407	1,830	1,577	3,357	1,803	1,554
1989	3,602	1,771	1,831	3,414	1,679	1,736
1990	3,247	1,911	1,336	2,990	1,760	1,230
1991	3,807	1,869	1,938	3,417	1,678	1,740
1992	3,051	1,577	1,474	2,610	1,349	1,261
1993	3,104	1,619	1,485	2,582	1,347	1,235
1994	2,518	1,123	1,395	2,052	915	1,137
1995	2,364	1,035	1,329	1,899	831	1,067
1996	2,016	635	1,381	1,591	501	1,090
1997	2,687	558	2,129	2,103	437	1,666
1998	2,262	496	1,766	1,760	386	1,374
1999	2,118	525	1,593	1,637	406	1,231
2000[r]	1,872	683	1,189	1,404	512	892
2001	1,479	382	1,097	1,082	279	802

Year	NET NEW ORDERS			BACKLOG AS OF DECEMBER 31		
	TOTAL	Military	Non-Military	TOTAL	Military	Non-Military
1987	$3,335	$1,995	$1,340	$3,849	$2,121	$1,728
1988	3,507	1,623	1,884	3,985	1,998	1,987
1989	6,113	2,475	3,638	6,410	2,595	3,815
1990	2,692	1,891	801	6,230	2,887	3,343
1991	5,661	1,087	4,574	8,422	2,327	6,095
1992	3,124	2,097	1,027	8,310	2,729	5,581
1993	1,708	710	998	6,543	1,903	4,640
1994	1,879	484	1,395	6,035	1,390	4,645
1995	2,805	444	2,361	6,630	1,065	5,565
1996	1,868	745	1,123	5,873	1,108	4,765
1997	2,009	477	1,532	5,568	1,023	4,545
1998	2,395	655	1,740	4,263	1,102	3,161
1999	3,896	687	3,209	6,182	1,017	5,165
2000[r]	1,425	493	932	5,499	816	4,683
2001	2,527	361	2,166	6,419	795	5,624

Source: Bureau of the Census, "Aerospace Industry (Orders, Sales, and Backlog)" (Annually).
  a Based on AIA's aerospace composite price deflator, 1987=100.

WORLDWIDE SPACE LAUNCHINGS[a]
WHICH ATTAINED EARTH ORBIT OR BEYOND
Calendar Years 1957–2001

Country	Total 1957–2001	1997	1998	1999	2000	2001[c]
TOTAL............................	5,179	76	80	74	82	49
U.S.S.R./C.I.S.	3,657	19	25	29	36	23
United States	1,233	37	36	30	29	18
European Space Agency	137	11	11	10	12	7
People's Republic of China ...	64	6	6	4	5	1
Japan	54	2	2	—	—	—
India..............................	11	1	—	1	—	—
Israel	3	—	—	—	—	—
Other[b]	20	—	—	—	—	—

Source: NASA, "Aeronautics and Space Report of the President" (Annually).
  a Number of launchings rather than spacecraft; some launches orbited multiple spacecraft.
  b Includes 10 by France, 8 by Italy (5 were U.S. spacecraft), 1 by Australia, and 1 by the United Kingdom.
  c Through September 30.

U.S. GOVERNMENT SPACECRAFT RECORD[a]
Calendar Years 1957–2001

Year	Earth Orbit[b] Success	Earth Orbit[b] Failure	Earth Escape[b] Success	Earth Escape[b] Failure	Year	Earth Orbit[b] Success	Earth Orbit[b] Failure	Earth Escape[b] Success	Earth Escape[b] Failure
1957	—	1	—	—	1982	21	—	—	—
1958	5	8	—	4	1983	31	—	—	—
1959	9	9	1	2	1984	35	3	—	—
1960	16	12	1	2	1985	37	1	—	—
1961	35	12	—	2	1986	11	4	—	—
1962	55	12	4	1	1987	9	1	—	—
1963	62	11	—	—	1988	16	1	—	—
1964	69	8	4	—	1989	24	—	2	—
1965	93	7	4	1	1990	40	—	1	—
1966	94	12	7	1[c]	1991	32[d]	—	—	—
1967	78	4	10	—	1992	26[d]	—	1	—
1968	61	15	3	—	1993	28[d]	1	1	—
1969	58	1	8	1	1994	31[d]	1	1	—
1970	36	1	3	—	1995	24[d]	2	1	—
1971	45	2	8	1	1996	30[d]	1	3	—
1972	33	2	8	—	1997	22	—	1	—
1973	23	2	3	—	1998	23	—	2	—
1974	27	2	1	—	1999	35	4	2	—
1975	30	4	4	—	2000	31	—	—	—
1976	33	—	1	—	2001[f]	16	—	3	—
1977	27	2	2	—	**TOTAL**	1,499	153	97	15
1978	34	2	7	—					
1979	18	—	—	—					
1980	16	4	—	—					
1981	20	1	—	—					

Source: NASA, "Aeronautics and Space Report of the President" (Annually).

a Payloads, rather than launchings; some launches account for multiple spacecraft. Includes spacecraft from cooperating countries launched on U.S. launch vehicles.

b The criterion of success is attainment of Earth orbit or Earth escape rather than judgement of mission success. "Escape" flights include all that were intended to go at least an altitude equal to the lunar distance from the Earth.

c This Earth-escape failure did attain Earth orbit and therefore is included in the Earth-orbit success totals.

d Excludes commercial satellites.

f Through September 30.

U.S. SPACE LAUNCH VEHICLES
As of 2001

Vehicle and Initial Launch & First Launch of This Modification	Stages	Thrust (Kilo-newtons)	Maximum Payload (Kg)[a]		
			185-Km Orbit	Geo-synch.-Transfer Orbit	Circular Sun-Synch. Orbit
Athena (1995)	1. Athena*	1,450.0	520	245	—
Atlas E (1958; 1968)	1. Atlas MA-3	1,739.5	820[b] 1,860[bc]	—	910[c]
Atlas I (1966; 1990)	1. Atlas MA-5 2. 2 Centaur I	1,952.0 146.8	—	2,255	—
Atlas II (1966; 1991)	1. Atlas MA-5A 2. 2 Centaur II	2,110.0 146.8	6,580 5,510[b]	2,810	4,300
Atlas IIA (1966; 1992)	1. Atlas MA-5A 2. 2 Centaur II	2,110.0 185.1	6,828 6,170[b]	3,062	4,750
Atlas IIAS (1966; 1993)	1. Atlas MA-5A plus 4 Castor IV* 2. 2 Centaur II	2,110.0 1,734.4 185.1	8,640 7,300[b]	3,606	5,800
Delta II 7900 Series (1960; 1990)	1. RS-270/A plus 9 Hercules GEM* 2. AJ10-118K 3. Star 48B*	1,043.0 4,388.4 42.4 66.4	5,089 3,890[b]	1,842[d]	3,175
Delta III (1998)[f]	1. RS-27 plus 9 Alliant GEM* 2. RL-10B 3. Star 48B*	1,043.0 5,479.2 110.0 66.4	8,292	3,810	6,768
Pegasus (1990)	1. Orion 50S* 2. Orion 50* 3. Orion 38*	484.9 118.2 31.9	380 280[b]	—	210
Pegasus XL (1994)[f]	1. Orion 50S-XL* 2. Orion 50-XL* 3. Orion 38*	743.3 201.5 31.9	460 350[b]	—	335
Space Shuttle (reusable) (1981)	0. 3 main engines (SSMEs) fire in parallel with solid-fueled rocket boosters (SRBs) 1. 2 SRBs mounted on external tank (ET) fire in parallel with SSMEs 2. 2 OMS	5,006.1 23,580.0 53.4	24,900[g]	5,900[h]	—
Taurus (1994)	0. Castor 120* 1. Orion 50S* 2. Orion 50* 3. Orion 38*	1,687.7 580.5 138.6 31.9	1,400 1,080[b]	255	1,020

(Continued on next page)

U.S. SPACE LAUNCH VEHICLES

As of 2001, continued

Vehicle and Initial Launch & First Launch of This Modification	Stages	Thrust (Kilo-newtons)	Maximum Payload (Kg)[a]		
			185-Km Orbit	Geo-synch.-Transfer Orbit	Circular Sun-Synch. Orbit
Titan II (1964; 1988)	1. 2 LR-87 2. LR-91	2,090.0 440.0	1,905[b]	—	—
Titan III (1964; 1989)	0. 2 5 1/2-segment, 3.05-m. dia* 1. 2 LR-87 2. LR-91	12,420.0 2,429.0 462.8	14,515	5,000[i]	—
Titan IV (1989)	0. 2 7-segment, 3.05-m. dia* 1. 2 LR-87 2. LR-91	14,000.0 2,429.0 462.8	17,700 14,110[b]	6,350[i]	—
Titan IV/Centaur (1994)	0. 2 7-segment, 4.3-m. dia* 1. 2 LR-87 2. LR-91 3. Centaur 4. SRMU	14,000.0 2,429.0 462.5 73.4 7,690.0	—	5,760	—

Source: NASA, "Aeronautics and Space Report of the President" (Annually).
* Solid propellant; all others are liquid.
a Due east launch except as indicated.
b Polar launch.
c With TE-M-364-4 upper stage.
d With Star 48B.
f First launch was a failure.
g In full performance configuration (280–420 km orbit).
h With IUS or TOS.
i With appropriate upper stage.

FEDERAL SPACE ACTIVITIES BUDGET AUTHORITY
Fiscal Years 1963–2001
(Millions of Dollars)

Year	TOTAL	NASA[a]	DoD	Commerce	Energy	Other[b]
1963	$ 5,433	$ 3,626	$ 1,550	$ 43	$214	$ —
1964	6,828	5,016	1,599	3	210	—
1965	6,953	5,138	1,574	12	229	—
1966	6,968	5,065	1,689	27	187	—
1967	6,707	4,830	1,664	29	184	—
1968	6,526	4,430	1,922	28	145	1
1969	6,005	3,822	2,013	20	118	32
1970	5,366	3,547	1,678	8	103	30
1971	4,775	3,101	1,512	27	95	40
1972	4,611	3,071	1,407	31	55	47
1973	4,863	3,093	1,623	40	54	53
1974	4,683	2,759	1,766	60	42	56
1975	4,965	2,915	1,892	64	30	64
1976	5,376	3,225	1,983	72	23	73
Tr.Qtr.	1,352	849	460	22	5	16
1977	6,046	3,440	2,412	91	22	81
1978	6,587	3,623	2,738	103	34	89
1979	7,314	4,030	3,036	98	59	91
1980	8,759	4,680	3,848	93	40	98
1981	10,054	4,992	4,828	87	41	106
1982	12,520	5,528	6,679	145	61	107
1983	15,674	6,328	9,019	178	39	110
1984	17,448	6,858	10,195	236	34	125
1985	20,277	6,925	12,768	423	34	127
1986	21,768	7,165	14,126	309	35	133
1987	26,562	9,809	16,287	278	48	140
1988	26,742	8,322	17,679	352	241	148
1989	28,563	10,097	17,906	301	97	162
1990	27,582	11,460	15,616	243	79	184
1991	27,999	13,046	14,181	251	251	270
1992	29,020	13,199	15,023	327	223	248
1993	27,901	13,064	14,106	324	165	242
1994	26,820	13,022	13,166	312	74	246
1995	23,946	12,543	10,644	352	60	347
1996	24,911	12,569	11,514	472	46	310
1997	24,973	12,457	11,727	448	35	306
1998	25,519	12,321	12,359	435	103	301
1999	26,644	12,459	13,203	575	105	302
2000	26,518	12,521	12,941	575	164	317
2001 [E]	28,703	13,304	14,326	576	145	352

Source: NASA, "Aeronautics and Space Report of the President" (Annually).
  a Excludes amounts for air transportation.
  b Departments of Interior, Transportation, and Agriculture, the National Science Foundation, and the Environmental Protection Agency.

FEDERAL SPACE ACTIVITIES OUTLAYS
Fiscal Years 1963–2001
(Millions of Dollars)

Year	TOTAL	NASA[a]	DoD	Commerce	Energy	Other[b]
1963	$ 4,079	$ 2,517	$ 1,368	$ 12	$181	$ 1
1964	5,930	4,131	1,564	12	220	3
1965	6,886	5,035	1,592	24	232	3
1966	7,719	5,858	1,637	28	188	7
1967	7,237	5,337	1,673	39	184	5
1968	6,667	4,595	1,890	29	147	6
1969	6,326	4,078	2,095	31	118	5
1970	5,453	3,565	1,756	24	103	5
1971	4,999	3,171	1,693	30	97	8
1972	4,772	3,195	1,470	37	60	10
1973	4,719	3,069	1,557	29	51	13
1974	4,854	2,960	1,777	64	39	14
1975	4,891	2,951	1,831	64	34	11
1976	5,314	3,336	1,864	71	26	16
Tr.Qtr.	1,361	869	458	23	8	4
1977	5,559	3,600	1,833	87	22	18
1978	6,188	3,582	2,457	101	29	20
1979	6,808	3,744	2,892	97	55	21
1980	7,734	4,340	3,162	89	49	94
1981	9,238	4,877	4,131	81	47	102
1982	10,542	5,463	4,772	142	60	106
1983	12,668	6,101	6,247	178	40	103
1984	14,813	6,461	8,000	209	33	109
1985	17,353	6,607	10,441	155	34	115
1986	18,683	6,756	11,449	317	35	127
1987	21,948	7,254	14,264	262	37	130
1988	23,521	8,451	14,397	334	199	140
1989	25,255	10,195	14,504	306	97	153
1990	25,788	12,292	12,962	279	79	177
1991	28,484	13,351	14,432	266	251	184
1992	27,998	12,838	14,437	298	223	202
1993	27,537	13,092	13,779	295	165	206
1994	23,929	12,363	10,973	297	83	213
1995	24,700	12,593	11,494	330	70	213
1996	24,675	12,694	11,353	354	46	228
1997	25,620	13,055	11,959	336	37	233
1998	25,827	12,866	12,230	326	97	308
1999	25,771	12,466	12,453	431	103	318
2000	26,633	12,427	13,207	517	165	317
2001 [E]	27,238	13,197	13,046	519	143	333

Source: NASA, "Aeronautics and Space Report of the President" (Annually).
  a Excludes amounts for air transportation.
  b Departments of Interior, Transportation, and Agriculture, the National Science Foundation, and the Environmental Protection Agency.

NATIONAL AERONAUTICS AND SPACE ADMINISTRATION
BUDGET AUTHORITY
Fiscal Years 1973–2003
(Millions of Dollars)

Year	TOTAL	Research and Development	Space Flight Control and Data Communications[a]	Construction of Facilities	Research & Program Management[b]
1973	$ 3,408	$2,599	$ —	$ 79	$ 730
1974	3,040	2,194	—	101	745
1975	3,231	2,323	—	143	765
1976	3,552	2,678	—	82	792
Tr.Qtr.	932	700	—	11	221
1977	3,819	2,856	—	118	845
1978	4,064	3,012	—	162	890
1979	4,559	3,477	—	148	934
1980	5,243	4,088	—	159	996
1981	5,522	4,334	—	117	1,071
1982	6,020	4,772	—	114	1,134
1983	6,875	5,539	—	139	1,197
1984	7,316	2,064[a]	3,772	223	1,256
1985	7,573	2,468	3,594	178	1,332
1986	7,807	2,619	3,670	176	1,342
1987	10,923	3,154	6,100	217	1,453
1988	9,062	3,280	3,806	213	1,763
1989	10,969	4,213	4,555	275	1,927
1990	12,324	5,225	4,645	218	2,023
1991	14,016	6,024	5,271	498	2,212
1992	14,317	6,848	5,352	525	1,576
1993	14,310	7,074	5,059	526	1,652
1994	14,570	7,534	4,835	493	1,708

Year	TOTAL	Science, Aeronautics, & Technology	Human Space Flight	Other[b]	Mission Support
1995[c]	$13,854	$5,936	$5,515	$(130)	$2,533
1996	13,886	5,929	5,457	17	2,483
1997	13,711	5,590	5,540	19	2,562
1998	13,649	5,690	5,560	19	2,380
1999	13,655	5,654	5,480	21	2,500
2000	13,602	5,582	5,488	21	2,511
2001	14,361	6,235	5,496	28	2,602
2002[dE]	15,013	8,114	6,873	26	—
2003[E]	15,118	8,918	6,173	27	—

Source: Office of Management and Budget, "Budget of the United States Government" (Annually).
    a Separate budget category beginning in 1984; funds formerly included under Research and Development.
    b Includes trust funds, Office of the Inspector General, National Space Grant Program, & GSA building delegation.
    c 1995 features major budget account restructuring.
    d Mission Support, as a separate category, discontinued; funds merged into other categories.

NATIONAL AERONAUTICS AND SPACE ADMINISTRATION OUTLAYS
Fiscal Years 1983–2003
(Millions of Dollars)

Year	TOTAL	Research and Development	Space Flight Control and Data Communications[a]	Construction of Facilities	Research & Program Management[b]
1983	$ 6,664	$5,316	$ —	$108	$1,240
1984	7,048	2,792[a]	2,915	109	1,232
1985	7,318	2,118	3,707	170	1,323
1986	7,404	2,615	3,267	189	1,332
1987	7,591	2,436	3,597	149	1,409
1988	9,092	2,916	4,362	166	1,648
1989	11,052	3,922	5,030	190	1,909
1990	12,429	5,094	5,117	218	2,000
1991	13,878	5,765	5,590	326	2,196
1992	13,961	6,579	5,118	463	1,802
1993	14,305	7,086	5,025	557	1,638
1994	13,695	6,758	4,899	371	1,666
1995[c]	5,098	3,286	1,409	305	98
1996[c]	1,022	510	241	265	6
1997[c]	317	101	92	122	2
1998[c]	138	40	34	64	—
1999[c]	47	18	2	27	—
2000[c]	31	18	1	12	—

Year	TOTAL	Science, Aeronautics, & Technology	Human Space Flight	Other[b]	Mission Support
1995[c]	$ 8,280	$2,708	$3,528	$ 15	$2,029
1996[c]	12,858	5,017	5,452	16	2,373
1997[c]	14,043	5,891	5,656	19	2,477
1998[c]	14,068	6,015	5,551	19	2,483
1999[c]	13,617	5,785	5,417	20	2,395
2000[c]	13,411	5,477	5,497	21	2,416
2001[d]	14,199	5,752	5,829	32	2,586
2002[Ef]	14,484	7,359	6,551	35	539
2003[Ef]	14,885	8,431	6,346	27	81

Source: Office of Management and Budget, "Budget of the United States Government" (Annually).
  a Separate budget category beginning in 1984; funds formerly included under Research and Development.
  b Includes trust funds, Office of Inspector General, National Space Grant Program, & GSA building delegation.
  c 1995 featured major budget account restructuring. Note: 1995–2000 outlays split between old and new account structure.
  d Continuing minimal outlays reported under old account structure included under Other beginning in 2001.
  f Mission Support, as a separate category, is being discontinued; funds merged into other categories.

NATIONAL AERONAUTICS AND SPACE ADMINISTRATION
BUDGET AUTHORITY BY MAJOR BUDGET ACCOUNT
FOR SELECTED PROGRAMS
Fiscal Years 2001–2003
(Millions of Dollars)

	2001	2002[E]	2003[E]
TOTAL	$14,253	$14,902	$15,000
HUMAN SPACE FLIGHT	$ 7,154	$ 6,830	$ 6,131
International Space Station	$ 2,128	$ 1,722	$ 1,492
Space Shuttle	3,119	3,273	3,208
Payload & ELV Support	90	91	88
Investments and Support	1,248	1,215	1,178
Space Communications & Data Systems	522	482	118
Safety, Mission Assurance, & Engineering	47	48	48
SCIENCE, AERONAUTICS, & TECHNOLOGY	$ 7,077	$ 8,048	$ 8,845
Space Science	$ 2,607	$ 2,867	$ 3,414
Biological & Physical Research	362	820	842
Earth Science	1,762	1,626	1,628
Aerospace Technology	2,213	2,508	2,816
Academic Programs	133	227	144
INSPECTOR GENERAL	$ 23	$ 24	$ 25

Source: "NASA Budget Briefing Background Material" (Annually).

DEPARTMENT OF DEFENSE SPACE PROGRAMS
PROCUREMENT[a] AND RDT&E
Fiscal Years 2001, 2002, and 2003
(Millions of Dollars)

Agency and Program	2001		2002 [E]		2003 [E]	
	Pro-cure-ment	RDT&E	Pro-cure-ment	RDT&E	Pro-cure-ment	RDT&E
AIR FORCE						
Defense Space Reconn Pgm	$ 8.9	$ 57.9	$ 6.8	$ 70.1	$390.7	$ 82.1
Defense Support Program...	102.0	12.8	109.0	6.1	114.4	2.1
DMSP	70.9	22.1	45.7	11.9	60.1	3.9
EELV..............................	286.3	377.6	98.0	315.3	158.9	57.6
Medium Launch Vehicles...	39.0	—	39.5	—	48.2	—
MilSat Com, EHF	24.6	229.8	15.9	479.7	45.7	825.8
MilSat Com, Polar	—	29.2	—	18.5	—	19.6
MILSAT Com, Transforma-tional Wideband	—	—	—	—	—	195.0
Milstar	—	224.6	—	228.7	—	148.9
NAVSTAR GPS	159.6	241.2	171.2	255.6	209.5	424.3
NPOESS	—	71.0	—	155.8	—	237.2
NUDET Detection System	17.0	14.9	18.9	21.5	—	21.2
Satellite Control Network ...	15.7	32.9	29.5	54.5	45.1	17.5
SBIRS-Low	—	233.5	—	—	—	—
SBIRS-High	—	550.1	—	438.7	—	814.9
Space-Based Laser	—	137.1	—	—	—	—
Space-Based Radar	—	—	—	24.8	—	47.9
Spacelift Range System	95.0	57.9	131.8	70.1	108.3	82.1
Titan Launch Vehicles	393.0	21.5	350.2	21.1	335.3	—
Wideband Gapfiller	24.7	119.1	371.0	99.0	189.7	20.0
ARMY						
DSCS[c]	$ 99.3	$ 9.5	$126.2	$ 13.2	$110.5	$ 12.2
NAVY						
Fleet Sat Comm	$ 94.7	$ NA	$ 77.2	$ NA	$ —	$ NA
MUOS	—	27.1	—	37.0	—	60.5
Sat Comm Systems	194.2	39.3	187.1	53.7	149.6	115.9

Source: Department of Defense, "Program Acquisition Costs by Weapon System" (Annually) and "Procurement Programs (P-1)" (Annually).
    a Total Obligational Authority for procurement including initial spares.
    b Air Force and BMDO funding.
    c Army and Air Force funding.
KEY: DMSP   = Defense Meteorological Satellite Program
        DSCS   = Defense Satellite Communications System
        EELV   = Evolved Expendable Launch Vehicle
        GPS   = Global Positioning System
        MUOS = Mobile User Objective System
        NPOESS = National Polar-orbiting Operational Environmental Satellite System
        SBIRS   = Space-Based InfraRed System

ORDERS AND BACKLOG OF COMMERCIAL GEOSYNCHRONOUS SATELLITES BY MANUFACTURER[a]

Calendar Years 1997–2001

	1997	1998	1999	2000	2001
ORDERS					
Value[b] (Millions of Dollars)	$3,351	$2,247	$1,847	$3,287	$2,380
Number of Satellites—TOTAL ...	29	19	16	31	25
Alcatel	6	1	2	6	2
Alenia Aerospazio	—	—	—	1	—
Astrium.............................	—	1	1	6	2
Boeing Satellite Systems.........	9	11	7	4	5
Israel Aircraft	—	—	—	—	1
Lockheed Martin Commercial Space Systems	5	4	3	2	5
NPO PM	—	—	—	3	—
Orbital	—	—	2	—	4
Space Systems Loral	9	2	1	9	6
BACKLOG[c]					
Value[b] (Milions of Dollars)	$4,973	$7,034	$7,884	$9,311	$9,235
Number of Satellites—TOTAL ...	44	62	67	80	84
Alcatel	6	8	9	12	13
Alenia Aerospazio	—	—	—	1	1
Astrium.............................	2	5	3	9	8
Boeing Satellite Systems.........	15	23	24	23	19
Israel Aircraft	—	—	—	—	1
Lockheed Martin Commercial Space Systems	7	11	13	10	10
NPO PM	—	—	—	3	3
Orbital	—	—	2	2	6
Space Systems Loral	14	15	16	20	23

Source: Futron Corporation.

    a   Excludes cancelled orders and orders on hold, without firm funding or business commitment, or with extended construction delay.

    b   Estimated using best available public information; where not available, used Futron estimates.

    c   Includes satellites on order during year and may include satellites that have been launched.

ORDERS AND BACKLOG OF COMMERCIAL LAUNCHES BY PROVIDER-COUNTRY

Calendar Years 1997–2001

	1997	1998	1999	2000	2001
ORDERS—TOTAL	32	20	28	38	35
China	4	2	1	—	—
India	—	—	—	—	—
Japan...............................	—	—	—	—	—
Russia	2	2	9	—	5
United States	9	2	6	17	15
Europe	17	14	12	16	13
Other multinational[a]	—	—	—	5	2
BACKLOG—TOTAL	146	140	141	129	163
China	4	6	7	7	7
India	—	—	—	—	—
Japan...............................	8	8	8	—	—
Russia	2	4	13	13	18
United States	65	58	55	47	68
Europe	49	46	40	46	53
Other multinational[a]	18	18	18	16	17

Source: Futron Corporation.
  a   Sea Launch.

AIR TRANSPORTATION

The world's airlines suffered a disastrous 2001, posting record losses and the second drop in traffic in aviation history. Operating profit declined by $22 billion from an $11 billion gain in 2000 to an $11 billion loss! Taking into account non-operating expenses such as interest and taxes, the industry's net loss totaled $12 billion, or -3.9% of revenue, down from a $3.7 billion gain. International Civil Aviation Organization statistics show that the airlines carried 35 million less passengers in 2001 and flew 54 million less passenger-miles. Load factors fell too from historic high levels, despite airlines quickly adjusting to lower traffic by grounding aircraft.

The impact of declining traffic, revenues, and profits fell squarely on U.S. airlines. Enplanements fell 6.6%, revenue ton-miles fell 6.4%, and operating revenue dropped 12% or $15 billion. Despite reacting immediately to match capacity to traffic and cutting costs, operating expenses increased 1.4% or $1.7 billion. For example, fuel prices remain near a 16-year high and fuel cost as a percentage of cash operating expenses fell from 14% to 13%, according to data from the Air Transport Association of America. Consequently, U.S. scheduled airline operating profits plummeted—falling $17 billion from a $7 billion gain in 2000 to a $10 billion loss in 2001. This marked the worst profit performance in history—ending an eight-year string of gains.

The world's airline fleet continued to grow in 2001, reaching 25,963 aircraft in service according to Air BP's "Turbine-Engined Fleets of the World's Airlines." There were more Boeing 737s in service than any other single type, with a total active fleet of 3,667. The McDonnell Douglas MD-80 was the next largest in 2001, numbering 1,141 aircraft. Comparing the active fleet in 2001 to the active fleet a year earlier reveals that the two dominant new-build narrowbodies, Boeing's 737 series and Airbus' A320 family (A319/A320/A321), grew more than any other aircraft type. The active A320 family population was 253 aircraft higher in 2001 at 1,607; the active 737 fleet increased by 236 aircraft.

The number of installed turbojet engines grew by 2,269 to 43,950 in 2001. Accounting for the growth: CFM International engines increased 797; "Other", 803; General Electric, 353; International Aero Engines, 236; and Rolls-Royce, 217. The number of Pratt & Whitney engines declined by 136 to 14,786.

Based on the latest available data, the number of hours flown by general aviation aircraft in the United States has decreased by 2.5% to 31 million in 2000. Similarly, the number of active general aviation aircraft has also decreased. However, since the enactment of the General Aviation Revitalization Act of 1994, the number of active single-engine aircraft has increased 22,300 to approximately 150,100.

OPERATING REVENUES, EXPENSES, AND RESULTS
OF WORLD SCHEDULED AIRLINES
Calendar Years 1998–2001
(Millions of Dollars)

	1998	1999	2000	2001[P]
OPERATING REVENUES:				
Scheduled Services:				
Passenger	$226,100	$231,410	$248,940	
Freight	29,420	30,400	33,840	
Mail	2,360	2,240	2,140	
				NA
Scheduled Services—Total	$257,880	$264,050	$284,920	
Non-Scheduled Services	9,660	11,440	11,610	
Incidental	27,960	30,010	31,970	
Operating Revenues—Total	$295,500	$305,500	$328,500	$305,300
OPERATING EXPENSES:				
Flight Operations	$ 75,080	$ 82,240	$ 98,790	
Maintenance & Overhaul	31,190	31,670	33,710	
Depreciation & Amortization	18,280	19,280	20,780	
User Charges & Station				NA
Expenses	50,010	51,760	54,720	
Passenger Services	29,770	31,520	31,780	
Ticketing, Sales & Promotion	40,110	40,130	40,450	
General, Administrative & Other	35,160	36,600	37,570	
Operating Expenses—Total	$279,600	$293,200	$317,800	$316,200
OPERATING RESULT	$ 15,900	$ 12,300	$ 10,700	($ 10,900)
Percent of Revenue	5.4%	4.0%	3.3%	–3.6%
NET RESULT[a]	$ 8,200	$ 8,500	$ 3,700	($ 12,000)
Percent of Revenue	2.8%	2.8%	1.1%	–3.9%

Source: International Civil Aviation Organization, "Civil Aviation Statistics of the World" (Annually).

a Net Result equals Operating Result minus non-operating items, including interest, income taxes, retirement of property and equipment, affiliated companies, and subsidies.

OPERATING REVENUES AND EXPENSES OF U.S. AIR CARRIERS[a]
DOMESTIC AND INTERNATIONAL OPERATIONS
Calendar Years 1967–2001
(Millions of Dollars)

Year	TOTAL			Domestic Operations			International Operations		
	Operating Revenues	Operating Expenses	Operating Profit (or Loss)	Operating Revenues	Operating Expenses	Operating Profit (or Loss)	Operating Revenues	Operating Expenses	Operating Profit (or Loss)
1967	$ 6,865	$ 6,157	$ 708	$ 4,981	$ 4,560	$ 421	$ 1,884	$ 1,597	$ 287
1968	7,753	7,248	505	5,691	5,397	295	2,062	1,852	210
1969	8,791	8,403	387	6,936	6,613	322	1,855	1,790	65
1970	9,290	9,247	43	7,180	7,181	(1)	2,109	2,066	44
1971	10,046	9,717	328	7,753	7,496	257	2,292	2,221	71
1972	11,163	10,578	584	8,652	8,158	493	2,512	2,420	91
1973	12,419	11,834	585	9,694	9,200	494	2,725	2,633	91
1974	14,703	13,978	725	11,546	10,761	785	3,157	3,218	(60)
1975	15,356	15,229	128	12,020	11,903	117	3,336	3,326	11
1976	17,503	16,781	721	13,899	13,324	575	3,605	3,457	147
1977	19,926	19,018	908	15,822	15,166	657	4,104	3,852	252
1978	22,892	21,527	1,366	18,189	17,172	1,018	4,703	4,355	348
1979	27,227	27,028	199	21,652	21,523	129	5,575	5,505	69
1980	33,728	33,949	(222)	26,404	26,409	(6)	6,543	6,766	(223)
1981	36,211	36,612	(401)	28,788	29,051	(264)	6,390	6,574	(184)
1982	36,066	36,804	(739)	28,728	29,478	(750)	6,435	6,452	(17)
1983	38,593	38,231	362	31,014	31,186	(171)	7,163	6,693	470
1984	44,060	41,946	2,114	35,394	33,812	1,582	7,975	7,485	490
1985	48,580	47,207	1,372	37,629	36,611	1,018	8,302	7,984	319
1986	50,086	48,855	1,231	41,001	39,984	1,060	8,621	8,458	163
1987	56,787	54,339	2,448	45,658	43,925	1,733	10,925	10,226	698
1988	63,679	60,236	3,443	50,187	47,739	2,448	13,402	12,403	998
1989	69,225	67,413	1,812	54,314	52,460	1,855	14,911	14,954	(43)
1990	75,984	77,898	(1,913)	57,994	58,983	(989)	17,990	18,914	(924)
1991	75,158	76,943	(1,785)	56,230	56,758	(528)	18,928	20,185	(1,257)
1992	78,140	80,585	(2,444)	57,654	58,801	(1,147)	20,486	21,784	(1,298)
1993	84,559	83,121	1,438	63,233	61,157	2,076	21,326	21,964	(637)
1994	88,313	85,600	2,713	65,949	63,758	2,191	22,364	21,842	522
1995	94,318	88,455	5,863	70,885	66,120	4,765	23,433	22,335	1,098
1996	101,937	95,728	6,209	76,891	71,573	5,317	25,047	24,155	892
1997	109,568	100,981	8,587	82,250	75,731	6,518	27,318	25,250	2,068
1998	113,465	104,137	9,328	86,494	78,389	8,105	26,971	25,749	1,223
1999	119,038	110,635	8,403	90,931	84,328	6,603	28,107	26,307	1,800
2000	130,299	123,314	6,985	98,896	93,579	5,317	31,403	29,736	1,668
2001 [p]	115,019	125,058	(10,040)	86,209	94,303	(8,094)	28,810	30,756	(1,946)

Source: Department of Transportation, Office of Airline Information, "Air Carrier Financial Statistics Quarterly" (Quarterly).
a Scheduled and non-scheduled service for all certificated route air carriers. Excludes supplemental air carriers, commuters, and air taxis.

SOURCES OF OPERATING REVENUES OF U.S. AIR CARRIERS[a]
DOMESTIC AND INTERNATIONAL OPERATIONS
Calendar Years 1987–2001
(Millions of Dollars)

Year	TOTAL	Passenger Service[b]	Mail	Freight[b] & Air Express	Excess Baggage	Other[c]
DOMESTIC OPERATIONS						
1987	$45,658	$37,492	$ 704	$4,952	$ 67	$ 2,443
1988	50,187	41,002	789	5,807	72	2,518
1989	54,314	43,670	767	5,408	70	4,399
1990	57,994	46,282	747	4,276	76	6,613
1991	56,230	44,594	734	4,487	78	6,337
1992	57,654	45,246	937	4,655	87	6,729
1993	63,233	49,289	974	5,266	91	7,612
1994	65,949	50,504	971	5,844	98	8,531
1995	70,885	53,971	1,050	6,546	92	9,227
1996	76,891	59,381	1,024	7,029	94	9,362
1997	82,250	62,549	1,087	7,497	99	11,017
1998	86,494	64,847	1,423	7,711	105	12,408
1999	90,931	67,777	1,475	8,053	118	13,509
2000	98,896	74,744	1,688	8,804	123	13,537
2001[p]	86,209	64,647	811	8,084	112	12,555
INTERNATIONAL OPERATIONS						
1987	$10,925	$ 8,374	$ 180	$1,783	$ 33	$ 555
1988	13,402	10,357	183	2,150	39	672
1989	14,911	11,181	188	2,417	47	1,078
1990	17,990	13,468	223	2,602	43	1,654
1991	18,928	14,103	223	3,134	50	1,419
1992	20,486	15,664	247	2,980	47	1,547
1993	21,326	15,915	237	3,220	49	1,905
1994	22,364	16,300	212	3,606	46	2,201
1995	23,433	16,788	216	3,994	48	2,387
1996	25,047	17,337	255	4,664	47	2,743
1997	27,318	18,320	275	5,156	56	3,511
1998	26,971	17,667	285	5,278	50	3,692
1999	28,107	18,011	264	5,921	46	3,865
2000	31,403	20,419	283	6,566	47	4,089
2001[p]	28,810	18,245	247	6,396	42	3,881

Source: Department of Transportation, Office of Airline Information, "Air Carrier Financial Statistics Quarterly" (Quarterly).
  a Scheduled and non-scheduled service for all certificated route air carriers. Excludes supplemental air carriers, commuters, and air taxis.
  b Scheduled and charter.
  c Includes subsidy, reservation cancellation fees, miscellaneous operating revenues, and other transport-related revenues.

OPERATING EXPENSES OF U.S. AIR CARRIERS[a]
DOMESTIC AND INTERNATIONAL OPERATIONS
Calendar Years 1987–2001
(Millions of Dollars)

Year	TOTAL	Flying Opera-tions	Mainte-nance	Passen-ger Service	Aircraft & Traffic Ser-vicing	Promo-tion and Sales	Depreci-ation & Amorti-zation	Other[b]
DOMESTIC OPERATIONS								
1987	$43,925	$12,509	$ 4,951	$4,169	$ 8,575	$ 7,399	$2,855	$ 3,468
1988	47,739	13,176	5,643	4,444	9,527	8,235	2,977	3,737
1989	52,460	14,749	6,184	4,775	9,449	8,718	3,078	5,507
1990	58,983	18,166	6,921	5,220	9,094	9,102	3,273	7,207
1991	56,758	16,831	6,682	5,068	9,140	8,856	3,217	6,964
1992	58,801	17,203	6,884	5,327	9,783	8,936	3,340	7,328
1993	61,157	17,622	7,025	5,241	10,172	9,387	3,621	8,089
1994	63,758	17,912	7,312	5,305	10,543	9,882	3,782	9,023
1995	66,120	18,926	7,656	5,281	11,103	9,974	3,762	9,417
1996	71,573	21,515	8,292	5,577	11,569	10,414	3,878	10,328
1997	75,731	22,156	9,475	5,854	12,058	10,780	3,940	11,469
1998	78,389	21,044	10,311	6,252	12,699	10,743	4,144	13,195
1999	84,328	22,820	11,161	6,763	13,796	10,760	4,657	14,372
2000	93,579	28,565	12,062	7,355	14,827	10,089	5,122	15,558
2001 [P]	94,303	27,796	12,093	7,217	15,373	8,935	6,188	16,701
INTERNATIONAL OPERATIONS								
1987	$10,226	$ 2,836	$ 1,096	$1,059	$ 1,749	$ 2,094	$ 533	$ 860
1988	12,403	3,230	1,332	1,280	2,193	2,742	618	1,009
1989	14,954	3,919	1,724	1,454	2,483	3,108	746	1,520
1990	18,878	5,454	2,051	1,738	2,657	3,833	887	2,295
1991	20,185	5,636	2,152	1,861	2,831	4,602	892	2,210
1992	21,784	5,843	2,148	2,204	3,255	5,229	1,033	2,073
1993	21,964	5,928	1,967	2,175	3,072	5,339	1,077	2,406
1994	21,842	5,842	2,064	2,311	3,336	4,335	1,237	2,716
1995	22,335	6,181	2,273	2,467	3,748	3,527	1,106	3,033
1996	24,155	7,279	2,616	2,596	3,736	3,354	1,483	3,091
1997	25,250	7,462	2,899	2,736	3,823	3,476	1,281	3,571
1998	25,749	7,158	2,955	2,920	3,978	3,374	1,438	3,926
1999	26,307	7,472	2,902	3,067	4,207	3,201	1,614	3,845
2000	29,736	9,504	3,093	3,211	4,565	3,282	1,751	4,329
2001 [P]	30,756	9,632	3,192	3,254	4,596	2,828	2,191	5,062

Source: Department of Transportation, Office of Airline Information, "Air Carrier Financial Statistics Quarterly" (Quarterly).
   a Scheduled and non-scheduled service for all certificated route air carriers. Excludes supplemental air carriers, commuters, and air taxis.
   b General and administrative and other transport-related expenses.

JET FUEL COSTS AND CONSUMPTION BY U.S. AIR CARRIERS[a]
Calendar Years 1978–2001

Year	Total Jet Fuel Cost (Millions of Dollars)	Gallons Consumed (Millions)	Cost Per Gallon (Cents)	Cost Index (1982 = 100)	Cost of Fuel as Percent of Cash Operating Expenses
1978	$ 4,069.6	10,359.5	39.3 ¢	39.0	19.4%
1979	6,354.0	11,042.0	57.5	57.1	24.3
1980	9,818.3	10,854.0	90.5	89.7	30.0
1981	10,827.5	10,326.9	104.8	104.0	29.9
1982	10,024.6	9,942.5	100.8	100.0	27.5
1983	9,320.9	10,472.5	89.0	88.3	24.7
1984	9,740.2	11,424.0	85.3	84.6	24.0
1985	9,689.8	12,072.6	80.3	79.6	22.3
1986	7,275.8	13,006.9	55.9	55.5	15.5
1987	7,895.6	14,139.6	55.8	55.4	15.0
1988	7,943.5	14,871.4	53.4	53.0	13.5
1989	9,104.3	15,115.8	60.2	59.7	13.9
1990	12,405.9	15,945.9	77.8	77.2	17.3
1991	10,275.2	14,682.9	70.0	69.4	14.5
1992	10,095.5	15,413.1	65.5	65.0	13.5
1993	9,378.7	15,569.3	60.2	59.7	12.4
1994	8,798.5	16,041.3	54.8	54.4	11.6
1995	9,053.2	16,233.1	55.8	55.3	12.0
1996	10,979.4	16,848.4	65.2	64.6	13.0
1997	10,990.1	17,450.8	63.0	62.5	12.5
1998	8,924.9	17,923.1	49.8	49.4	9.9
1999	9,526.0	18,366.9	51.9	51.4	10.0
2000	14,876.0	18,854.6	78.9	78.3	14.0
2001	13,914.7	17,717.8	78.5	77.9	12.9

Source: Air Transport Association of America, "Airline Cost Index" (Quarterly).
   a   Majors and Nationals.

TOTAL ASSETS AND INVESTMENT IN EQUIPMENT BY U.S. AIR CARRIERS
Calendar Years 1969–2001
(Millions of Dollars)

Year	Total Assets	Value of Flight Equipment	Value of Ground Property & Equipment & Other[a]	Less: Reserves for Depreciation & Overhaul	Equals: Net Value of Owned Operating Property & Equipment	Investment in Operating Property and Equipment as a Percent of Total Assets
1969	$ 12,069	$ 9,943	$ 1,516	$ 3,560	$ 7,899	65.4%
1970	12,913	10,950	1,951	4,120	8,782	68.0
1971	12,998	11,221	2,028	4,649	8,600	66.2
1972	13,635	11,918	2,225	5,115	9,028	66.2
1973	14,464	12,908	2,424	5,693	9,639	66.6
1974	15,200	13,538	2,539	6,252	9,826	64.6
1975	15,064	14,035	2,635	6,823	9,847	65.4
1976	15,454	14,399	2,792	7,585	9,605	62.2
1977	16,869	14,822	2,997	8,141	9,679	57.4
1978	20,745	16,127	3,367	8,799	10,696	51.6
1979	24,907	18,561	3,985	9,746	12,800	51.4
1980	28,900	20,859	4,682	10,309	15,233	52.7
1981	30,513	22,375	5,175	11,028	16,521	54.1
1982	31,525	23,786	5,424	11,405	17,804	56.5
1983	35,213	26,588	6,191	12,910	19,868	56.4
1984	36,769	28,509	6,061	14,043	20,527	55.8
1985	40,978	30,402	6,772	15,467	21,707	53.0
1986	47,105	31,750	8,468	14,764	25,454	54.0
1987	51,436	33,177	9,223	15,580	26,820	52.1
1988	56,047	35,781	10,248	17,450	28,579	51.0
1989	62,454	38,812	11,903	19,018	31,697	50.8
1990	67,769	40,215	13,523	20,593	33,144	48.9
1991	70,332	42,897	14,285	22,009	35,173	50.0
1992	75,426	48,563	15,219	24,445	39,337	52.2
1993	82,399	51,513	15,438	24,949	42,003	51.0
1994	84,442	51,951	15,844	26,476	41,319	48.9
1995	89,782	56,018	16,804	29,056	43,766	48.7
1996	95,184	59,206	16,661	30,029	45,838	48.2
1997	105,226	66,523	17,643	32,789	51,377	48.8
1998	118,308	75,385	19,980	35,992	59,373	50.2
1999	133,711	86,269	21,826	39,060	69,035	51.6
2000	146,300	98,404	22,095	41,880	78,620	53.7
2001[P]	158,417	102,791	23,061	42,405	83,447	52.7

Source: Department of Transportation, Office of Airline Information, "Air Carrier Financial Statistics Quarterly" (Quarterly).
   a Includes land and construction in progress.

TRAFFIC STATISTICS
WORLD AIRLINE SCHEDULED SERVICE[a]
Calendar Years 1970–2001

Year	Passengers Carried	Freight Tons Carried	Passenger-Miles Performed	Seat-Miles Available	Passenger Load Factor	Total (Passengers & Baggage, Freight, Mail)	Freight	Mail
						Ton-Miles Performed		
	(Millions)		(Billions)		(Percent)	(Billions)		
1970	383	6.7	286	522	55%	38.81	8.23	2.10
1971	411	7.4	307	568	54	41.42	9.06	1.99
1972	450	8.0	348	609	57	46.69	10.29	1.90
1973	489	9.1	384	667	58	51.91	12.01	1.97
1974	514	9.5	408	688	59	55.27	13.03[r]	1.98
1975	534	9.6	433	733	59	58.08[r]	13.27	1.99
1976	576	10.3	475	789	60	63.88	14.75	2.08
1977	610	11.1	508	837	61	68.79	16.19[r]	2.17
1978	679	11.7	582	902	65	77.77	17.77	2.24
1979	754	12.1	659	999	66	86.89	19.19	2.35
1980	748	12.2	677	1,071	63	89.71	20.12	2.52
1981	752	12.0	695	1,091	64	92.81[r]	21.15	2.60
1982	766	12.8	710	1,115	64	94.84[r]	21.60	2.65
1983	798	13.5	739	1,151	64	100.27	24.05	2.74
1984	848	14.8	794	1,226	65	109.05	27.17	2.95
1985	899	15.1	850	1,293	66	114.86	27.29	3.01
1986	960	16.2	902	1,389	65	122.47	29.58	3.11
1987	1,028	17.7	988	1,471	67	134.57	33.10	3.22
1988	1,082	19.0	1,060	1,568	68	145.28	36.48	3.31
1989	1,109	19.9	1,102	1,621	68	152.73[r]	39.14	3.46
1990	1,165	20.3	1,177	1,740	68	161.11	40.27	3.65
1991	1,135	19.3	1,147	1,727	66	158.03	40.11	3.48
1992	1,146	19.5	1,199	1,821	66	165.85	42.90	3.51
1993	1,142	19.9	1,211	1,872	65	171.67[r]	46.88	3.58
1994	1,233	22.6	1,305	1,969	66	187.28	52.89	3.71
1995	1,304	24.5	1,397	2,087	67	201.33[r]	56.94	3.86
1996	1,391	25.6	1,511	2,214	68	217.24[r]	61.10	3.97
1997	1,457	29.1	1,599	2,316	69	235.75	70.47	4.10
1998	1,471	29.2	1,633	2,385	68	238.77	69.74	3.94
1999	1,562	31.0	1,739	2,517	69	253.72[r]	74.42	3.92
2000	1,656	33.3	1,875	2,646	71	274.78	80.80	4.14
2001[p]	1,621	31.7	1,821	2,639	69	263.96	75.81	3.62

Source: International Civil Aviation Organization (ICAO).
  a Includes international and domestic traffic on scheduled service performed by the airlines of the 185 states which were members of ICAO in 2001.

TRAFFIC STATISTICS
U.S. AIR CARRIER SCHEDULED SERVICE[a]
Calendar Years 1967–2001

Year	Revenue Ton-Miles (Millions)			Total Available Ton-Miles (Millions)	Total Revenue Load Factor (Percent)	Aircraft Revenue Miles (Millions)	Average Overall Flight Stage Length (Miles)	Average Available Seats per Aircraft Mile
	TOTAL	Passenger	Cargo[b]					
1967	13,036	9,561	3,475	26,968	48.3%	1,834	371	101
1968	15,249	11,023	4,226	33,221	45.9	2,146	401	107
1969	16,898	12,197	4,701	38,664	43.7	2,385	443	112
1970	18,166	13,171	4,994	41,693	43.6	2,426	473	117
1971	18,685	13,565	5,120	44,139	42.3	2,378	476	125
1972	20,746	15,241	5,506	45,583	45.5	2,376	471	129
1973	22,242	16,196	6,046	49,019	45.4	2,448	477	135
1974	22,425	16,292	6,133	46,848	47.9	2,258	478	140
1975	22,186	16,281	5,905	47,254	46.9	2,241	476	143
1976	24,121	17,899	6,222	49,325	48.9	2,320	480	146
1977	25,909	19,322	6,587	52,284	49.6	2,419	490	149
1978	29,679	22,678	7,001	54,765	54.2	2,520	502	152
1979	33,390	26,202	7,189	60,844	54.9	2,791	517	154
1980	32,603	25,519	7,084	62,983	51.8	2,816	526	158
1981	31,949	24,889	7,060	61,186	52.2	2,703	519	161
1982	32,850	25,964	6,886	62,401	52.6	2,699	544	167
1983	35,756	28,183	7,573	65,385	54.7	2,809	558	169
1984	38,697	30,512	8,185	72,223	53.6	3,134	575	168
1985	41,329	33,640	7,689	76,059	54.3	3,320	569	168
1986	45,681	36,655	9,026	85,140	53.7	3,725	580	168
1987	50,469	40,453	10,016	92,209	54.7	3,988	606	167
1988	53,800	42,330	11,469	97,899	55.0	4,141	618	169
1989	55,458	43,271	12,187	100,082	55.4	4,193	633	169
1990	58,342	45,793	12,549	107,559	54.2	4,491	649	170
1991	56,925	44,795	12,130	105,599	53.9	4,416	651	169
1992	61,054	47,855	13,199	112,749	54.2	4,661	661	169
1993	63,088	48,968	14,120	115,473	54.6	4,846	669	166
1994	67,989	51,938	16,052	120,798	56.3	5,033	668	163
1995	70,987	54,066	16,921	126,154	56.3	5,293	657	160
1996	75,621	57,866	17,754	131,381	57.6	5,501	668	160
1997	80,852	60,342	20,510	137,544	58.8	5,659	696	160
1998	82,304	61,809	20,496	141,722	58.1	5,838	704	159
1999	86,817	65,205	21,613	149,561	58.0	6,168	715	157
2000	93,163	69,276	23,888	159,441	58.4	6,574	728	154
2001[p]	87,164	65,166	21,997	157,744	55.3	6,514	741	152

Source: Department of Transportation, Office of Airline Information, "Air Carrier Traffic Statistics Monthly" (Monthly).
  a   Includes international and domestic operations.
  b   Includes freight, air express, U.S. and foreign mail.

PASSENGER STATISTICS
U.S. AIR CARRIER SCHEDULED SERVICE
DOMESTIC AND INTERNATIONAL OPERATIONS
Calendar Years 1987–2001

Year	Revenue Passenger Enplanements (Thousands)	Average Passenger Trip-Length (Miles)	Revenue Passenger Miles (Millions)	Available Seat Miles (Millions)	Revenue Passenger Load Factor[a]
DOMESTIC OPERATIONS					
1987	416,831	779	324,637	526,958	61.6
1988	419,210	786	329,309	536,663	61.4
1989	416,331	793	329,975	530,079	62.3
1990	423,565	803	340,231	563,065	60.4
1991	412,360	806	332,566	543,638	61.2
1992	431,693	806	347,931	557,989	62.4
1993	443,172	799	354,177	571,489	62.0
1994	481,755	787	378,990	585,438	64.7
1995	499,000	791	394,708	603,917	65.4
1996	530,708	802	425,596	626,389	67.9
1997	542,001	817	442,640	640,319	69.1
1998	559,653	812	454,430	649,362	70.0
1999	582,880	824	480,134	687,502	69.8
2000	610,601	833	508,403	714,454	71.2
2001[P]	570,128	843	480,313	695,178	69.1
INTERNATIONAL OPERATIONS					
1987	30,847	2,588	79,834	121,763	65.6
1988	35,404	2,655	93,992	140,140	67.1
1989	37,361	2,750	102,739	154,297	66.6
1990	41,995	2,803	117,695	170,310	69.1
1991	39,941	2,889	115,389	171,561	67.3
1992	43,415	3,009	130,622	194,784	67.1
1993	45,348	2,988	135,508	200,151	67.7
1994	47,093	2,981	140,391	198,893	70.6
1995	48,773	2,992	145,948	203,160	71.8
1996	50,526	3,029	153,067	208,682	73.3
1997	52,724	3,049	160,779	216,913	74.1
1998	53,232	3,074	163,656	224,728	72.8
1999	53,079	3,239	171,913	230,917	74.4
2000	55,549	3,319	184,354	242,496	76.0
2001[P]	52,003	3,295	171,350	235,308	72.8

Source: Department of Transportation, Office of Airline Information, "Air Carrier Traffic Statistics Monthly" (Monthly).
   a   Revenue passenger miles as a percent of available seat miles.

AIR CARGO STATISTICS
U.S. COMMERCIAL AIR CARRIERS
Fiscal Years 1974–2001
(Millions of Revenue-Ton-Miles)

Year	TOTAL	Freight/Express		Mail	
		Domestic	International	Domestic	International
1974	6,590.5	3,009.4	2,431.3	687.0	462.8
1975	6,138.8	2,779.0	2,240.1	687.4	432.3
1976	6,473.8	2,907.6	2,472.6	694.8	398.8
1977	6,705.0	3,092.5	2,460.7	748.2	403.6
1978	7,354.6	3,484.2	2,713.1	797.7	359.7
1979	7,540.2	3,590.8	2,761.7	830.2	357.4
1980	7,602.5	3,419.2	2,892.9	919.8	370.6
1981	7,362.0	3,365.3	2,651.6	991.7	353.4
1982	7,297.1	3,144.0	2,792.3	996.0	364.9
1983	8,129.9	3,808.8	2,910.2	1,036.6	374.3
1984	9,269.2	4,390.7	3,328.0	1,142.8	407.6
1985	8,902.7	3,941.8	3,340.4	1,201.8	418.6
1986	10,510.9	4,872.0	3,995.4	1,232.3	411.2
1987	12,288.3	5,781.7	4,780.6	1,314.0	412.0
1988	14,260.5	6,699.3	5,701.6	1,422.2	437.3
1989	16,081.0	7,412.8	6,749.0	1,462.2	457.1
1990	16,283.2	7,532.5	6,770.3	1,477.5	502.9
1991	16,327.4	7,451.0	6,907.0	1,462.4	507.0
1992	16,792.9	7,861.8	6,819.0	1,612.5	499.5
1993	18,420.2	8,557.4	7,541.0	1,816.7	505.1
1994	20,789.9	9,334.5	8,957.2	1,988.8	509.4
1995	23,227.6	10,342.1	10,278.0	2,073.6	533.0
1996	24,216.9	10,655.3	10,874.6	2,126.4	560.6
1997	26,952.3	11,177.9	12,926.4	2,276.2	571.8
1998	28,350.3	11,527.3	13,992.0	2,300.8	529.3
1999	28,101.8	11,527.1	13,617.3	2,447.8	509.6
2000	30,056.6	12,157.2	14,825.7	2,541.6	532.1
2001 E	28,462.3	11,778.9	14,021.4	2,155.0	507.0

Source: Federal Aviation Administration, Office of Aviation Policy & Plans.

TURBINE-ENGINED AIRCRAFT IN THE WORLD AIRLINE FLEET BY MODEL
Calendar Years 1997–2001

	1997	1998	1999	2000	2001
TOTAL.................................	22,110	23,002	24,128	25,173	25,963
Turbojets—TOTAL.................	14,024	14,621	15,453	16,405	17,291
Aerospatiale SE-210 Caravelle	12	12	12	9	5
Aerospatiale Corvette	—	—	2	2	2
Airbus A300........................	397	383	389	408	396
Airbus A310........................	224	227	220	217	216
Airbus A319........................	66	118	206	309	398
Airbus A320........................	612	685	773	880	994
Airbus A321........................	72	109	143	165	215
Airbus A330........................	63	87	131	170	201
Airbus A340........................	119	138	161	181	200
Antonov 72/74	8	9	9	15	19
Antonov 124	16	18	17	17	17
Antonov 225	1	1	1	1	1
Avro RJ-70/85/100	100	120	138	158	163
B.Ae./Aerospatiale Concorde	13	13	13	12	12
B.Ae. 146...........................	208	206	203	206	195
B.Ae. One-Eleven	122	105	101	97	88
B.Ae. (HS) 125	18	17	11	13	15
Beech 400 Beechjet	3	3	3	3	3
Boeing 707/720	112	98	94	86	76
Boeing 717	—	—	11	44	92
Boeing 727	1,322	1,263	1,189	1,144	1,077
Boeing 737	2,752	2,968	3,176	3,431	3,667
Boeing 747	1,040	1,042	1,041	1,027	1,029
Boeing 757	770	818	879	924	952
Boeing 767	663	710	752	799	816
Boeing 777	111	174	261	313	373
Canadair CL-601 Challenger...	2	2	2	1	1
Canadair Regional Jet............	189	258	351	451	559
Cessna Citation I/II/III	41	35	34	41	39
Dassault Falcon 10/20/50	60	61	71	69	73
Embraer ERJ-135/140/145	34	91	186	345	497
Fairchild 328 Jet	—	—	11	43	67
Fokker F-28 Fellowship	184	160	169	172	148
Fokker 70	36	42	42	42	42
Fokker 100	274	278	277	271	257
Gulfstream II/III/IV G-1159 ...	16	14	14	14	12
Ilyushin IL-62	105	86	83	85	105
Ilyushin IL-76	227	215	226	257	285
Ilyushin IL-86	80	75	79	81	87
Ilyushin IL-96	7	7	8	8	9
Israel Aircraft 1121/1124	11	18	10	8	8
Learjet	53	49	85	86	85
Lockheed L-1011 Tristar	169	156	135	117	102
Lockheed L-1329 Jetstar.........	1	1	1	1	1
MBB Hansa HFB-320............	16	17	9	9	6
McDonnell Douglas DC-8......	257	261	225	227	200
McDonnell Douglas DC-9......	759	749	714	676	627
McDonnell Douglas DC-10 ...	345	354	340	319	317
McDonnell Douglas MD-11 ...	171	180	187	188	191

(Continued on next page)

TURBINE-ENGINED AIRCRAFT IN THE WORLD AIRLINE FLEET BY MODEL
Calendar Years 1997–2001, continued

	1997	1998	1999	2000	2001
Turbojets (continued)					
McDonnell Douglas MD-80	1,142	1,154	1,180	1,164	1,141
McDonnell Douglas MD-90	62	97	109	110	106
Tupolev Tu-134	189	189	188	193	202
Tupolev Tu-154	451	438	446	439	482
Tupolev Tu-204	6	7	9	11	14
Yakolev Yak-40/42	313	303	326	356	406
Turbine-Powered					
Helicopters—TOTAL	1,014	1,371	1,449	1,438	1,429
Aerospatiale SA-315 Lama	2	2	2	—	—
Aerospatiale SA-316 Alouette III	5	5	5	5	2
Aerospatiale SA-318 Alouette II	1	1	1	1	1
Aerospatiale (Nurtanio) SA-330 Puma	20	18	22	19	3
Aerospatiale AS-332 Super Puma	78	73	72	74	92
Aerospatiale AS-350 Ecureuil/ Astar	104	112	125	124	124
Aerospatiale AS-355 Ecureuil 2/ Twinstar.............................	15	23	28	26	23
Aerospatiale SA-365 Dauphin II	26	26	36	43	45
Agusta A109	—	—	—	—	2
Bell (Agusta/Fuji) 204	4	2	4	4	4
Bell 205	14	11	15	14	14
Bell 206 Jetranger/Longranger ...	151	361	363	353	323
Bell 212	101	112	120	115	115
Bell 214	7	17	14	13	11
Bell 222 UT..........................	2	3	2	2	1
Bell 230	—	3	—	—	—
Bell 407	1	44	53	54	74
Bell 412	31	55	56	59	71
Boeing 107	15	15	15	15	15
Boeing Vertol BV-234	9	9	9	9	9
Eurocopter EC-120	—	—	—	2	2
Hughes (Kawasaki) 500/369D ...	17	16	13	12	9
Kamov Ka-26	16	16	19	16	16
Kamov Ka-32	2	—	—	—	—
MBB BK-117	2	13	10	8	7
MBB/Nurtanio Bo.105	67	103	101	100	106
Mil Mi-2	40	32	28	26	26
Mil Mi-6	6	6	6	6	—
Mil Mi-8	91	83	107	102	96
Mil Mi-14............................	1	1	1	1	1
Mil Mi-26............................	7	7	5	8	8
Sikorsky S-55T.......................	6	6	—	1	—
Sikorsky S-58T.......................	2	1	1	—	1
Sikorsky S-61	90	65	79	81	85
Sikorsky S-62	1	1	—	—	—
Sikorsky S-64	5	5	22	24	5
Sikorsky S-76	75	124	115	121	138

(Continued on next page)

TURBINE-ENGINED AIRCRAFT IN THE WORLD AIRLINE FLEET BY MODEL
Calendar Years 1997–2001, continued

	1997	1998	1999	2000	2001
Turboprops—TOTAL..................	7,072	7,010	7,226	7,330	7,243
Aerospatiale N.262/Mohawk 298	9	11	12	12	14
Aerospatiale/Aeritalia ATR 42 ...	296	299	296	311	306
Aerospatiale/Aeritalia ATR 72 ...	177	202	222	223	255
Airtech CN-235	24	24	33	23	23
Antonov An-8	2	—	6	4	6
Antonov An-12	71	83	81	114	112
Antonov An-22	3	3	1	1	1
Antonov An-24/26/30/32	530	499	475	526	582
B.Ae. ATP.............................	50	57	55	49	28
B.Ae. Viscount......................	18	12	12	12	—
B.Ae. (HP-137) Jetstream 31	287	233	258	223	190
B.Ae. Jetstream 41	91	92	92	93	90
B.Ae. HS-748	125	124	118	109	110
Beech 18 Turbo	20	18	9	9	8
Beech 90 King Air	46	39	46	39	39
Beech 99	138	139	110	142	146
Beech 100 King Air	39	39	47	51	48
Beech 200/300 Super King Air...	122	111	112	131	134
Beech 1300	9	6	9	8	7
Beech 1900C/D	430	467	469	490	397
Canadair CL-44	—	—	4	3	3
CASA/Nurtanio C-212 Aviocar ...	113	105	110	109	108
Cessna 208 Caravan I	608	601	647	661	705
Cessna F406 Caravan II	30	31	30	29	33
Cessna 425/441 Conquest I/II ...	14	19	19	15	15
Convair 580/600/640	107	107	106	104	103
DHC-2/3 Turbo Beaver/Otter ...	20	20	24	28	35
DHC-5 Buffalo	1	1	1	1	1
DHC-6 Twin Otter	395	371	365	349	345
DHC-7 Dash 7	69	71	69	67	68
DHC-8 Dash 8	424	444	489	523	557
Dornier DO-228	114	118	121	103	105
Dornier DO-328	61	73	83	91	89
Douglas DC-3T Turbo Express ...	1	1	3	3	3
Embraer EMB-110 Bandeirante...	200	199	199	181	181
Embraer EMB-120 Brasilia.........	308	316	307	300	290
Embraer EMB-121 Xingu	2	2	3	1	1
Fokker/Fairchild F-27/FH-227 ...					
Friendship	318	278	276	257	251
Fokker 50.............................	171	167	188	184	170
GAF Nomad	15	15	16	12	12
Grumman G-73 Turbo Mallard...	5	5	6	6	6
Grumman G-159 Gulfstream I ...	30	27	27	17	14

(Continued on next page)

TURBINE-ENGINED AIRCRAFT IN THE WORLD AIRLINE FLEET BY MODEL
Calendar Years 1997–2001, continued

	1997	1998	1999	2000	2001
Turboprops (continued)					
Handley Page Herald	2	1	1	1	1
Harbin YU-12 II	42	48	48	46	47
IAI Arava	3	3	4	3	2
Ilyushin IL-18	34	32	41	38	44
Ilyushin IL-114	2	2	3	3	3
LET L-410	115	118	141	178	186
Lockheed L-188 Electra	36	44	43	40	35
Lockheed L-100/L-382 Hercules	45	35	44	43	41
Mitsubishi MU-2B	15	16	21	18	22
Nihon AMC YS-11	63	49	46	37	33
Piaggio P-180 Avanti	—	—	—	—	1
Pilatus Britten-Norman BN-2T					
Turbo Islander	6	6	5	5	4
Pilatus PC-6 Turbo Porter	30	24	23	27	25
Pilatus PC-XII	2	14	21	22	22
Piper PA-31T/42 Cheyenne ...	20	20	22	19	18
Piper T-1040	14	13	13	9	9
PZL (Antonov) An-28	3	3	27	43	42
Rockwell Turbo Commander	11	9	8	8	9
Saab SF-340A/B	396	432	414	447	380
Saab 2000	42	45	43	48	48
Shorts SC-5 Belfast	2	2	2	2	1
Shorts SC-7 Skyliner/Skyvan ...	32	30	27	26	25
Shorts 330	48	42	37	39	39
Shorts 360	103	93	102	101	106
Swearingen Merlin	53	55	58	50	52
Swearingen Metro	394	379	398	392	385
Transall C-160	—	—	6	6	6
Xian (Antonov) Y-7	66	66	65	65	66
TOTAL AIRCRAFT IN SERVICE	22,110	23,002	24,128	25,173	25,963
Number Manufactured in U.S.	12,487	13,139	13,537	13,857	13,988
Percent Manufactured in U.S.	56.5%	57.1%	56.1%	55.0%	53.9%
Turbojet Aircraft in Service	14,024	14,621	15,453	16,405	17,291
Number Manufactured in U.S.	9,789	10,126	10,430	10,714	10,906
Percent Manufactured in U.S.	69.8%	69.3%	67.5%	65.3%	63.1%
Turboprop Aircraft in Service ...	7,072	7,010	7,226	7,330	7,243
Number Manufactured in U.S.	2,172	2,165	2,226	2,266	2,207
Percent Manufactured in U.S.	30.7%	30.9%	30.8%	30.9%	30.5%
Turbine-Powered Helicopters					
In Service	1,014	1,371	1,449	1,438	1,429
Number Manufactured in U.S.	526	848	881	877	875
Percent Manufactured in U.S.	51.9%	61.9%	60.8%	61.0%	61.2%

Source: Air BP Lubricants, "Turbine-Engined Fleets of the World's Airlines," compiled by Aviation Data Service, Inc. (Annually).
NOTE: "Turbine-Engined Fleets of the World's Airlines" covers aircraft in airline service as of December 31. Excludes air taxi operators.

NUMBER AND PERCENT OF CIVIL TURBOJET ENGINES
IN WORLD AIRLINE FLEET BY MANUFACTURER AND AIRCRAFT MODEL
As of December 2001

Aircraft Manufacturer and Model	Total Installed Engines	Engine Manufacturers					
		P&W	GE	RR	CFM	IAE	Other
TOTAL ENGINES	43,950	14,786	6,355	4,385	8,966	1,486	7,972
PERCENT OF TOTAL ...	100.0%	33.6%	14.5%	10.0%	20.4%	3.4%	18.1%
Airbus A300[a]	542	20%	80%	−%	−%	−%	−%
Airbus A300B4-600R ...	318	54	47	−	−	−	−
Airbus A310[a]	162	36	64	−	−	−	−
Airbus A310-300	288	43	57	−	−	−	−
Airbus A319	786	−	−	−	68	32	−
Airbus A320[a]	36	−	−	−	100	−	−
Airbus A320-200	1,948	−	−	−	59	42	−
Airbus A321[a]	178	−	−	−	47	53	−
Airbus A321-200	246	−	−	−	58	42	−
Airbus A330[a]	186	44	8	48	−	−	−
Airbus A330-300	222	45	14	41	−	−	−
Airbus A340[a]	312	−	−	−	100	−	−
Airbus A340-300X	512	−	−	−	100	−	−
Antonov AN-72	12	−	−	−	−	−	100
Antonov AN-74	30	−	−	−	−	−	100
Antonov AN-124	72	−	−	−	−	−	100
AS Corvette..................	6	100	−	−	−	−	−
AS Caravelle	24	67	−	33	−	−	−
AS/BAe Concorde	48	−	−	100	−	−	−
Avro Int'l RJ	656	−	−	−	−	−	100
BAe 1-11.....................	206	−	−	100	−	−	−
BAe 146	812	−	−	−	−	−	100
BAe/HS 125	38	5	−	11	−	−	84
Beech 400 Beechjet	18	100	−	−	−	−	−
Boeing B-707[a]	84	100	−	−	−	−	−
Boeing B-707-320C	404	100	−	−	−	−	−
Boeing B-717	186	−	−	100	−	−	−
Boeing B-720	4	100	−	−	−	−	−
Boeing B-727 series[a]	1,257	88	−	12	−	−	−
Boeing B-727-200 Adv F	672	100	−	−	−	−	−
Boeing B-727-200 ADV	1,596	100	−	−	−	−	−
Boeing B-737[a]	566	61	−	−	39	−	−
Boeing B-737-200 ADV	1,346	100	−	−	−	−	−
Boeing B-737-300	2,122	−	−	−	100	−	−
Boeing B-737-400	944	−	−	−	100	−	−
Boeing B-737-500	768	−	−	−	100	−	−
Boeing B-737-700	678	−	−	−	100	−	−
Boeing B-737-800	1,098	−	−	−	100	−	−
Boeing B-747[a]	2,136	48	41	11	−	−	−
Boeing B-747-200B	592	70	14	16	−	−	−
Boeing B-747-400	1,684	40	35	25	−	−	−
Boeing B-757[a]	332	23	−	77	−	−	−
Boeing B-757-200	1,602	45	−	55	−	−	−
Boeing B-767[a]	298	16	84	−	−	−	−
Boeing B-767-200ER	276	49	51	−	−	−	−
Boeing B-767-300	202	25	75	−	−	−	−
Boeing B-767-300ER	896	37	57	7	−	−	−

(Continued on next page)

NUMBER AND PERCENT OF CIVIL TURBOJET ENGINES
IN WORLD AIRLINE FLEET BY MANUFACTURER AND AIRCRAFT MODEL
As of December 2001, continued

Boeing B-777[a]	230	54%	8%	36%	–%	–%	–%
Boeing B-777-200ER	520	21	38	41	–	–	–
Canadair CL 600/601	6	–	67	–	–	–	33
Canadair Regional Jet series[a]	40	–	100	–	–	–	–
Canadair Regional Jet[b]	466	–	100	–	–	–	–
Canadair Regional Jet 200	644	–	100	–	–	–	–
Cessna 500s	92	96	–	4	–	–	–
Cessna 650	16	–	–	–	–	–	100
Dassault Falcon	158	–	82	–	–	–	18
Embraer ERJ-135	176	–	–	–	–	–	100
Embraer ERJ-140	44	–	–	–	–	–	100
Embraer ERJ-145	776	–	–	–	–	–	100
Fairchild Dornier 328 Jet	140	100	–	–	–	–	–
Fokker F-28	314	–	–	100	–	–	–
Fokker 70	84	–	–	100	–	–	–
Fokker 100	532	–	–	100	–	–	–
Gulfstream II/III/IV	30	–	–	100	–	–	–
IAI 1124/1125	18	–	–	–	–	–	100
Ilyushin IL-62[a]	44	–	–	–	–	–	100
Ilyushin IL-62M	404	–	–	–	–	–	100
Ilyushin IL-76[a]	104	–	–	–	–	–	100
Ilyushin IL-76MD	448	–	–	–	–	–	100
Ilyushin IL-76TD	516	–	–	–	–	–	100
Ilyushin IL-86	336	–	–	–	–	–	100
Ilyushin IL-96	36	–	–	–	–	–	100
Learjet 23/24/25	124	–	100	–	–	–	–
Learjet 35/36/55/60	106	6	–	–	–	–	94
Lockheed JetStar	4	–	–	–	–	–	100
Lockheed L-1011	384	–	–	100	–	–	–
MBB HFB-320 Hansa Jet	18	–	100	–	–	–	–
Douglas DC-8	944	60	–	–	40	–	–
Douglas DC-9[a]	506	100	–	–	–	–	–
Douglas DC-9-30	846	100	–	–	–	–	–
Douglas DC-10[a]	645	18	82	–	–	–	–
Douglas DC-10-30	327	–	100	–	–	–	–
MDC MD-11 series[a]	261	22	78	–	–	–	–
MDC MD-11[b]	324	53	47	–	–	–	–
MDC MD-80s[a]	324	100	–	–	–	–	–
MDC MD-82	1,176	100	–	–	–	–	–
MDC MD-83	530	100	–	–	–	–	–
MDC MD-88	314	100	–	–	–	–	–
MDC MD-90-30	230	–	–	–	–	100	–
Rockwell Sabreliner	2	–	100	–	–	–	–
Tupolev TU-134[a]	198	–	–	–	–	–	100
Tupolev TU-134A3	244	–	–	–	–	–	100
Tupolev TU-154[a]	189	–	–	–	–	–	100
Tupolev TU-154B2	591	–	–	–	–	–	100
Tupolev TU-154M	729	–	–	–	–	–	100
Tupolev TU-204	22	–	–	36	–	–	64
Yakovlev YAK-40	894	–	–	–	–	–	100
Yakovlev YAK-42	329	–	–	–	–	–	100

Source: Aerospace Industries Association, based on data from Aviation Data Service.
  a Data for major (100 or more aircraft) series excluded and reported separately.
  b Series bearing same designation as model number, but qualifies for separate reporting as a major series.
KEY: AS = Aerospatiale; BAe = British Aerospace; CFM = CFM International; GE = General Electric;
  IAE = International Aero Engines; IAI = Israel Aircraft Industries; MBB = Messerschmitt Bolkow Blohm;

ACTIVE[a] U.S. AIR CARRIER FLEET
BY TYPE OF AIRCRAFT, NUMBER OF ENGINES, AND MODEL
As of December 1997–2001

	1997	1998	1999	2000[r]	2001
TOTAL......................................	7,616	8,111	8,228	8,055	8,497
Turbojets—TOTAL.........................	5,108	5,411	5,630	5,956	6,296
Four-Engine—TOTAL	450	447	441	432	419
Boeing 707	3	—	1	—	1
Boeing 747	201	201	188	190	209
B.Ae./AVRO 146......................	26	18	46	54	54
McDonnell Douglas DC-8............	220	228	206	188	155
Three-Engine—TOTAL	1,224	1,238	1,181	1,061	996
Boeing 727	874	882	811	720	674
Lockheed L-1011	79	70	66	52	24
McDonnell Douglas DC-10/MD-11 ...	271	286	304	289	298
Twin-Engine—TOTAL	3,434	3,726	4,008	4,463	4,881
Airbus A-300	68	61	68	85	113
Airbus A-310	28	39	39	39	46
Airbus A-319	2	23	40	111	170
Airbus A-320	119	143	162	207	231
Airbus A-321	—	—	—	—	23
Airbus A-330	—	—	—	5	9
BAe HS-125.............................	—	—	1	—	—
Beech 400	—	1	1	—	1
Boeing 717	—	—	2	22	57
Boeing 737	1,077	1,080	1,179	1,280	1,359
Boeing 757	487	510	555	585	599
Boeing 767	234	261	278	315	340
Boeing 777	23	36	53	89	96
Canadair CL-600......................	77	152	187	251	324
Cessna C500/C501	—	10	9	—	—
Dassau AMD	—	27	27	—	26
Embraer ERJ-135	—	—	7	39	69
Embraer ERJ-145	11	55	95	151	217
Fokker F-28	142	147	145	145	133
Israel Aircraft 1124....................	—	1	1	—	1
Learjet LR-25	3	7	8	6	2
Learjet LR-31	—	1	1	—	—
Learjet LR-35	9	11	11	11	2
McDonnell Douglas DC-9/MD-80/ MD-90	1,154	1,158	1,133	1,122	1,060
Mitsubishi MU-300....................	—	2	5	—	2
North American NA-265	—	1	1	—	1
Turboprops—TOTAL.....................	1,646	1,832	1,788	1,475	1,494
Four-Engine—TOTAL	45	39	28	29	24
De Havilland DHC-7	5	7	6	7	4
Lockheed 188 Electra.................	22	17	14	14	12
Lockheed 382	18	15	8	8	8

(Continued on next page)

ACTIVE[a] U.S. AIR CARRIER FLEET
BY TYPE OF AIRCRAFT, NUMBER OF ENGINES, AND MODEL
As of December 1997–2001, continued

	1997	1998	1999	2000[r]	2001
Twin-Engine—TOTAL	1,596	1,789	1,759	1,440	1,470
Airtech CN-235	—	—	1	1	1
Beech BE90	2	8	6	1	3
Beech BE99	28	36	38	9	14
Beech BE100	1	2	4	3	3
Beech BE200	7	19	19	7	8
Beech BE300	—	—	—	—	2
Beech BE1900	243	325	239	224	185
B.Ae. ATP	9	—	9	—	—
B.Ae. Jetstream	215	203	184	145	100
CASA C212 Aviocar	—	3	4	4	4
Cessna CE208B	—	137	167	44	159
Cessna C441	2	4	2	1	—
Convair 580/600/640	19	15	12	10	5
Curtis C-46	—	—	—	1	2
DeHavilland DHC-6	49	54	54	38	51
DeHavilland DHC-8	154	169	180	183	190
Dornier DO328	47	35	39	65	85
Embraer EMB110	1	1	1	1	1
Embraer EMB120	227	218	225	197	176
Fairchild/Fokker F-27/FH-227	44	38	38	38	38
Grumman G-73	5	5	3	5	5
Gulfstream 690A	1	—	—	—	—
Gulfstream GA1159	—	—	—	—	3
Mitsubishi MU-2	11	13	14	13	—
Piper PA31T	10	6	6	4	4
Piper 42	2	2	2	—	—
Saab-Fairchild SF340	253	271	275	272	255
Shorts SC-7	3	3	3	3	3
Shorts SD-3	33	15	20	10	13
SNAIS ATR-42	95	83	79	69	66
SNAIS ATR-72	55	60	60	70	65
Swearingen SA-226	7	4	3	1	2
Swearingen SA-227	73	60	72	27	27
Single-Engine—TOTAL	5	4	1	—	—
Piston-Engine—TOTAL	728	751	688	585	580
Four-Engine—TOTAL	19	17	19	17	16
Douglas DC-6	19	17	19	17	16
Three-Engine—TOTAL	4	3	3	3	3
Pilatus Britten-Norman BN2A-MK-3 Turbo Islander	4	3	3	3	3
Twin-Engine—TOTAL	298	391	292	255	173
Single-Engine—TOTAL	407	340	374	310	388
Helicopters—TOTAL	134	117	122	39	127

Source: Federal Aviation Administration.

NOTE: Effective 1978, includes certificated route air carriers, supplemental air carriers (charters), multi-engine aircraft in passenger service of commuters, and all aircraft over 12,500 pounds operated by Part 121 and Part 135 commuter operators.

a "Active aircraft" equals the average number of aircraft reported in operation during the last quarter of the year.

ACTIVE U.S. MILITARY AIRCRAFT[a]
Fiscal Years 1980–2001

Year	TOTAL[a]	Fixed-Wing Aircraft				Helicopters
		TOTAL	Jet	Turboprop	Piston	
1980	18,969	11,362	8,794	1,869	699	7,607
1981	19,363	11,645	9,111	1,943	591	7,718
1982	21,728	12,063	9,647	1,900	516	9,665
1983	18,652	11,603	9,495	1,745	363	7,049
1984	18,833	11,661	9,551	1,777	333	7,172
1985	19,333	11,929	9,640	1,881	408	7,404
1986	20,157	11,919	9,730	1,803	386	8,238
1987	20,514	12,054	9,819	1,865	370	8,460
1988	21,010	12,481	9,954	2,222	305	8,529
1989	19,223	11,893	9,501	2,131	261	7,330
1990	20,017	12,817	10,360	2,199	258	7,200
1991	19,966	12,587	10,221	2,119	247	7,379
1992	19,210	11,936	9,672	2,035	229	7,274
1993	17,231	9,681	7,651	1,852	178	7,550
1994[E]	17,018	9,803	7,786	1,835	182	7,215
1995[E]	16,207	9,277	7,294	1,754	229	6,930
1996[b]	20,554	10,154	7,798	2,199	157	10,400
1997	20,245	9,677	7,364	2,151	162	10,568
1998	15,585	9,187[c]	7,082	1,951	120	6,398
1999	16,062	9,015[d]	6,981	1,908	115	7,047
2000	15,902	8,777[d]	6,738	2,023	5	7,125
2001	16,155	9,024[f]	6,790	2,084	118	7,131

Source: Aerospace Industries Association.
  a Includes Army, Air Force, Navy, and Marine regular service aircraft, as well as Reserve and National Guard Aircraft.
  b Prior years data provided by Office of the Secretary of Defense and limited to aircraft in the continental United States.
  c Includes 34 gliders.
  d Includes 11 gliders.
  f Includes 32 gliders.

ACTIVE U.S. CIVIL AIRCRAFT[a]
As of December 31, 1966–2000
(Thousands)

Year	TOTAL	Air Carrier[b]	General Aviation Aircraft					
				Fixed-Wing Aircraft			Rotor-craft[c]	Other[d]
			TOTAL	Multi-Engine	Single-Engine			
					4-place & over	3-place & less		
1966	107.0	2.272	104.7	13.5	53.0	35.7	1.6	0.9
1967	116.6	2.452	114.2	14.7	56.9	39.7	1.9	1.1
1968	126.8	2.586	124.2	16.8	61.0	42.8	2.4	1.3
1969	133.5	2.690	130.8	18.1	63.7	45.0	2.6	1.4
1970	134.4	2.679	131.7	18.3	64.8	44.9	2.3	1.6
1971	133.8	2.642	131.1	17.9	64.5	44.8	2.4	1.7
1972	147.6	2.583	145.0	19.8	71.0	49.4	2.8	1.9
1973	156.1	2.599	153.5	21.9	74.8	51.4	3.1	2.3
1974	164.0	2.472	161.5	23.4	78.9	53.0	3.6	2.5
1975	171.0	2.495	168.5	24.6	82.6	54.4	4.1	2.8
1976	180.8	2.492	178.3	25.7	88.2	56.7	4.5	3.2
1977	186.8	2.473	184.3	26.7	92.0	57.3	4.7	3.6
1978	201.3	2.545	198.8	28.8	101.5	59.2	5.3	4.0
1979	213.9	3.609	210.3	31.3	106.0	62.4	5.9	4.8
1980	214.9	3.808	211.0	31.7	107.9	60.5	6.0	4.9
1981	217.2	3.973	213.2	33.3	108.0	59.9	7.0	5.0
1982	213.9	4.027	209.8	34.2	106.5	57.7	6.2	6.2
1983	217.5	4.203	213.3	34.6	107.1	59.1	6.5	5.9
1984	225.3	4.370	220.9	35.6	109.9	62.0	7.1	6.3
1985	201.2	4.678	196.5	31.3	98.5	54.9	6.0	5.8
1986	210.2	4.909	205.3	32.0	102.0	58.3	6.5	6.5
1987	208.0	5.253	202.7	30.8	100.4	59.3	5.9	6.3
1988	201.9	5.660	196.2	30.1	98.1	55.6	6.0	6.4
1989	210.8	5.778	205.0	31.9	100.5	58.4	7.0	7.2
1990	204.1	6.083	198.0	30.6	97.6	56.4	6.9	6.6
1991	202.9	6.054	196.9	29.7	97.8	55.1	6.2	8.1
1992	193.0	7.320	185.7	26.8	91.6	53.2	6.0	8.0
1993	184.4	7.297	177.1	22.8	91.6	42.5	4.7	15.5
1994	180.3	7.370	172.9	22.3	87.3	40.5	4.7	18.1
1995	190.0	7.411	188.1	24.6	93.6	44.1	5.8	19.9
1996	194.8	7.478	191.1	25.6	93.8	44.3	6.6	20.9
1997	200.0	7.616	192.4	26.2	95.0	45.7	6.8	18.8
1998	212.8	8.111	204.7	31.0	102.5	41.8	7.4	22.1
1999	227.7	8.228	219.5	32.8	109.3	42.6	7.4	27.3
2000	225.6	8.055	217.5	33.2	108.0	42.1	7.2	27.1

Source: Federal Aviation Administration.
a "Active aircraft" must have a current U.S. registration and have flown during the calendar year. Prior to 1971, only a current U.S. registration was necessary.
b Effective 1978, includes certificated route air carriers, supplemental air carriers (charters), multi-engine aircraft in commuter passenger service, and all aircraft over 12,500 pounds operated by air taxis, commercial operators, and travel clubs.
c Includes autogiros; excludes air carrier helicopters.
d Includes gliders, dirigibles, balloons, and experimental aircraft.

U.S. GENERAL AVIATION[a]
TYPE OF AIRCRAFT AND HOURS FLOWN
Calendar Years 1996–2000

	1996	1997	1998	1999	2000
NUMBER OF ACTIVE AIRCRAFT BY TYPE (Thousands)					
All Aircraft—TOTAL	191.1	192.4	204.7	219.5	217.5
Fixed-Wing:	163.7	166.9	175.2	184.7	183.3
Piston:	153.6	156.1	163.0	171.9	170.5
Single-Engine	137.4	140.0	144.2	150.9	149.4
Twin-Engine	16.1	15.9	18.7	20.9	21.0
Other	0.1	0.1	0.1	0.1	0.1
Turboprop:	5.7	5.6	6.2	5.7	5.8
Twin-Engine	4.9	4.9	5.1	4.6	5.0
Other	0.8	0.7	1.1	1.0[r]	0.7
Turbojet:	4.4	5.2	6.1	7.1	7.0
Twin-Engine	4.1	4.6	5.5	6.4	6.2
Other	0.3	0.5	0.6	0.7	0.8
Rotorcraft:	6.6	6.8	7.4	7.4	7.2
Piston	2.5	2.3	2.5	2.6	2.7
Turbine	4.1	4.5	4.9	4.9	4.5
Balloons, Dirigibles, and Gliders...	4.2	4.1	5.6	6.8	6.7
Experimental	16.6	14.7	16.5	20.5	20.4
HOURS FLOWN BY TYPE OF AIRCRAFT (Thousands)					
All Aircraft—TOTAL	26,909	27,713	28,100	31,756	30,975
Fixed-Wing: Piston	20,091	20,743	20,402	22,895	22,199
Turboprop	1,768	1,655	1,765	1,811[r]	2,031
Turbojet	1,543	1,713	2,226	2,738	2,755
Rotorcraft: Piston	591	343	430	556	531
Turbine	1,531	1,739	1,912	2,188	1,777
Balloons, Dirigibles, and Gliders...	227	192	295	318[r]	374
Experimental	1,158	1,327	1,071	1,247[r]	1,307
AVERAGE HOURS FLOWN ANNUALLY BY TYPE					
All Aircraft—TOTAL	140.8	144.0	137.3	144.7	142.4
Fixed-Wing: Piston	130.8	132.9	125.2	133.2	130.2
Turboprop	309.3	294.5	285.8	319.0	352.5
Turbojet	348.7	330.7	367.0	384.6	393.5
Rotorcraft: Piston	235.9	152.2	169.0	217.0	198.1
Turbine	376.9	384.3	391.8	448.0	397.6
Balloons, Dirigibles, and Gliders...	53.6	46.8	52.8	47.1	55.8
Experimental	69.6	90.4	64.9	60.8	64.0

Source: Federal Aviation Administration.
   a Beginning in 1993, commuters were excluded from the survey.

U.S. GENERAL AVIATION
ACTIVE AIRCRAFT AND HOURS FLOWN BY PRIMARY USE[a]
Calendar Years 1996–2000

Primary Use	1996	1997	1998	1999	2000
ACTIVE AIRCRAFT AS OF DECEMBER 31 (Thousands)					
TOTAL......................	191.1	192.4	204.7	219.5	217.5
Executive....................	9.9	10.4	11.3	10.8	11.0
Business	30.7	27.7	32.6	24.5	25.2
Air Taxi[b]	4.1	4.8	4.9	4.3	3.7
Instructional	12.7	14.7	11.4	16.1	14.9
Personal	113.4	115.6	124.3	147.1	148.2
Aerial Application	5.0	4.9	4.6	4.3	4.3
Aerial Observation.........	3.0	3.3	3.2	3.2	5.1
Aerial Other	NA	NA	NA	0.4	0.1
Sight Seeing	0.7	0.7	0.7	0.8	0.9
Public Use	4.5	4.1	4.0	4.1	NA
Air Tours	0.1	0.2	0.3	0.3	0.3
External Load	0.4	0.2	0.3	0.2	0.2
Medical	NA	NA	NA	0.8	0.9
Other Work	1.0	0.7	1.1	2.4	1.8
Other	5.6	5.3	6.0	NA[r]	NA
HOURS FLOWN (Thousands)					
TOTAL......................	26,909	27,713	28,100	31,756	30,975
Executive....................	2,898	2,878	3,213	3,616	3,458
Business	3,259	3,006	3,523	3,598	3,670
Air Taxi[b]	1,734	2,008	2,400	1,897	1,550
Instructional	4,759	4,956	3,961	5,893	5,369
Personal	9,037	9,644	9,781	11,294	11,699
Aerial Application	1,713	1,562	1,306	1,415	1,401
Aerial Observation.........	1,057	1,261	812	1,243	1,632
Aerial Other	NA	NA	NA	120	233
Sight Seeing	195	127	169	220	198
Air Tours	100	114	183	146	646
Public Use	1,047	1,096	1,373	1,111	NA
Medical	NA	NA	NA	461	442
External Load	191	112	153	128	171
Other Work	265	139	286	613	506
Other	656	819	940	NA[r]	NA

Source: Federal Aviation Administration, "General Aviation and Air Taxi Activity Survey" (Annually).
a Definitions of "primary use" categories available in Glossary of "FAA Statistical Handbook."
b Air taxis under 12,500 pounds.

ACTIVE U.S. CIVIL AIRCRAFT
BY PRIMARY USE AND TYPE OF AIRCRAFT
As of December 31, 2000

Primary Use	TOTAL	Fixed-Wing			Rotor-craft[a]	Other[b]
		Turbojet	Turboprop	Piston		
TOTAL............................	225,588	12,957	7,237	171,098	7,189	27,107
Air Carrier—TOTAL	8,055	5,956	1,475	585	39	—
Large	7,368	5,939	1,380	49	—	—
Small	687	17	95	536	39	—
General Aviation—TOTAL	217,533	7,001	5,762	170,513	7,150	27,107
Executive........................	11,003	5,078	2,831	2,352	578	165
Business	25,169	466	1,145	22,740	342	476
Air Taxi[c]	3,686	649	536	2,042	424	35
Instructional	14,883	33	21	13,271	725	831
Personal	148,192	526	520	121,471	1,262	24,412
Aerial Application	4,294	170	367	3,174	513	71
Aerial Observation............	5,093	21	69	3,255	1,691	56
Aerial Other	1,022	—	143	530	323	26
Sight Seeing	881	—	4	236	117	522
Air Tours	333	—	—	122	166	31
External Load	234	—	—	—	221	13
Medical	930	24	76	219	570	41
Other Work	1,787	33	37	1,084	211	424

Source: Federal Aviation Administration.
NOTE: Detail may not add to totals because of estimating procedures.
   a   Includes helicopters and autogiros.
   b   Includes gliders, dirigibles, balloons, and experimental aircraft.
   c   Limited to Air taxis under 12,500 pounds. Otherwise, aircraft included in "Air Carrier."

U.S. CIVIL AND JOINT-USE AIRCRAFT FACILITIES[a]
BY STATE AND BY TYPE
As of December 31, 2001

State	Total[a]	Public[b]	Paved	Lighted	State	Total[a]	Public[b]	Paved	Lighted
Alabama.........	245	95	156	103	Nevada	125	52	66	36
Alaska............	591	405	72	173	New Hampshire	112	27	53	22
Arizona	302	83	170	87	New Jersey	372	54	158	56
Arkansas.........	297	100	178	111	New Mexico......	173	63	81	55
California	938	264	664	281	New York	572	164	222	150
Colorado	418	79	180	85	North Carolina ...	372	115	160	121
Connecticut ...	151	24	92	31	North Dakota ...	432	92	85	92
Delaware	45	11	17	15	Ohio	732	174	289	181
Dist. of Col. ...	16	3	15	4	Oklahoma.........	435	149	216	134
Florida	818	131	359	179	Oregon	436	99	168	80
Georgia	430	109	194	121	Pennsylvania......	780	137	324	141
Hawaii	49	15	39	16	Rhode Island......	28	10	20	8
Idaho	238	120	83	51	South Carolina ...	178	66	85	74
Illinois............	874	121	304	182	South Dakota ...	181	75	77	77
Indiana	618	115	170	117	Tennessee	282	83	156	91
Iowa	317	122	176	136	Texas	1,804	387	861	451
Kansas	408	143	142	129	Utah	138	47	88	48
Kentucky	200	63	117	67	Vermont............	80	17	19	11
Louisiana	479	82	258	87	Virginia	409	67	165	90
Maine............	159	65	53	33	Washington	472	137	227	136
Maryland	228	38	85	49	West Virginia ...	116	38	69	36
Massachusetts	224	44	120	42	Wisconsin	545	133	191	149
Michigan	474	232	201	179	Wyoming	118	41	57	41
Minnesota	508	158	161	146	50 States—TOTAL	19,252	5,279	8,429	5,119
Mississippi	235	83	120	87	Puerto Rico	44	11	36	12
Missouri	541	134	238	144	Virgin Islands ...	7	2	3	2
Montana.........	254	121	107	91	S. Pacific[c]	18	10	12	7
Nebraska	303	92	121	93	TOTAL	19,321	5,302	8,480	5,141

FACILITIES BY CLASS

Class	TOTAL[a]	Public[b]	Private
Airports	13,663	5,025	8,638
Heliports	5,106	79	5,027
Seaplane Bases	466	195	271
Stolports	86	3	83
TOTAL...	19,321	5,302	14,019

Source: Federal Aviation Administration.
   a   Included in these data are facilities having joint civil-military use.
   b   "Public" refers to use, whether publicly or privately owned.
   c   American Samoa, Guam, and Trust Territories.

HELIPORTS/HELIPADS[a] IN THE UNITED STATES BY STATE
As of 2001

State	TOTAL Helipads in State	Private Use		Public Use	
		Heliports & Helistops	Helipads at Airports	Heliports & Helistops	Helipads at Airports
Alabama	77	76	—	—	1
Alaska	35	24	1	7	3
Arizona	114	106	1	—	7
Arkansas	82	79	—	—	3
California	414	392	2	—	20
Colorado	169	165	—	—	4
Connecticut	93	88	—	2	3
Delaware	13	12	—	1	—
District of Columbia	18	17	—	1	—
Florida	280	275	2	1	2
Georgia	108	106	—	—	2
Hawaii	18	16	—	—	2
Idaho	38	36	1	—	1
Illinois	255	246	4	5	—
Indiana	119	112	4	2	1
Iowa	88	87	—	—	1
Kansas	42	37	1	—	4
Kentucky	58	58	—	—	—
Louisiana	242	234	2	4	2
Maine	19	18	—	—	1
Maryland	67	64	1	—	2
Massachusetts	131	127	—	2	2
Michigan	88	84	1	3	—
Minnesota	53	51	—	—	2
Mississippi	48	47	—	1	—
Missouri	131	128	1	1	1
Montana	30	27	—	2	1
Nebraska	36	34	1	—	1
Nevada	32	29	—	—	3
New Hampshire	54	53	—	—	1

(Continued on next page)

HELIPORTS/HELIPADS[a] IN THE UNITED STATES BY STATE

As of 2001, continued

State	TOTAL Helipads in State	Private Use		Public Use	
		Heliports & Helistops	Helipads at Airports	Heliports & Helistops	Helipads at Airports
New Jersey	248	242	—	3	3
New Mexico	26	24	1	1	—
New York	159	146	1	7	5
North Carolina	72	69	—	3	—
North Dakota	17	16	—	—	1
Ohio	203	191	1	10	2
Oklahoma	92	87	—	5	—
Oregon	101	96	3	2	—
Pennsylvania	307	296	1	8	2
Rhode Island	16	15	—	1	—
South Carolina	30	28	—	—	2
South Dakota	26	26	—	—	—
Tennessee	90	87	1	1	1
Texas	445	431	3	5	6
Utah	45	43	—	—	2
Vermont	19	19	—	—	—
Virginia	130	126	—	—	4
Washington	138	130	3	1	4
West Virginia	36	33	—	—	3
Wisconsin	86	85	—	—	1
Wyoming	26	24	—	—	2
TOTAL	5,264	5,041	36	79	108

Source: Helicopter Association International, "2002 Helicopter Annual" (Annually).
NOTE: 96.4 percent of all U.S. helicopter landing areas are private, while 3.6 percent are public.
a Excludes temporary heliports, offshore heliports, and infrequently used helicopter landing sites.

RESEARCH AND DEVELOPMENT

The National Science Foundation (NSF) estimates that the United States funded a total of $265 billion of R&D in 2000—the latest figures available. Industry is shouldering an increasing share of R&D spending. In 1997, industry's share of total R&D spending was 64%. That figure has crept up a percentage point or so each year; and in 2000, the $181 billion spent by industry on R&D accounted for some 68% of the total. At the same time, though the federal government's spending is rising in absolute terms, its percentage of the total is declining. In 1997, the federal government supplied a little less than 31% of the R&D dollars in the United States. By 2000, the federal government's $70 billion accounted for less than 27%. At $200 billion, industry also performed the majority of R&D, according to the NSF's Annual Survey of Industrial R&D. Colleges and universities combined were the next-largest R&D performer in 2000—accounting for $30 billion of the year's R&D dollars. The federal government accounted for $19 billion.

Funding for aerospace industry-performed R&D declined $4.1 billion, or 28%, in 2000—again using the latest figures available from the NSF. Federal funding of aerospace industry R&D totaled $6.4 billion—down from $9.1 billion; and company funding declined $1.4 billion to $3.9 billion. This marks the lowest level in two decades before adjustment for inflation and even longer after. Coincident with this decline in funding, the number of R&D-performing scientists and engineers (R&D S&Es) employed by the aerospace industry has dropped from 144,800 in 1986 to just 25,100 in 2001. Similarly, the percentage of the nation's R&D S&Es employed in the aerospace industry has fallen from 21.6% to 2.4%.

AIA has been at the forefront of this crisis and the U.S. government has responded. In FY 2001, federal outlays for R&D (total, not just aerospace) increased $12 billion after languishing for six years at the same real spending level. DoD outlays increased $7.2 billion to $45

billion while NASA R&D spending increased 9% to $7 billion and Energy R&D rose 5%. The DoD continued to be the government's largest single spender on R&D, accounting for more than half of all federal funding. Other agencies, such as the NSF, the National Institutes of Health, and the Transportation and Agriculture Departments saw their collective R&D outlays rise 19% to $27 billion. Further, in FY 2002, federally-funded R&D is scheduled to increase $11 billion to $97 billion. DoD R&D is due to increase $4.8 billion and Energy R&D will jump 24% or $1.6 billion. NASA's R&D funding, on the other hand, was cut $383 million.

Ballistic Missile Defense continues to dominate DoD's RDT&E account. BMD funding totaled $4.2 billion in FY 2001 and is scheduled to increase to $7.0 billion in 2002. Other DoD aerospace programs receiving the greatest levels of RDT&E funding in FY 2001 included: F-22 Raptor, $1.4 billion; Joint Strike Fighter, $682 million; RAH-66 Comanche, $591 million; and SBIRS-High, $550 million.

TOTAL U.S. FUNDS FOR RESEARCH AND DEVELOPMENT
BY SOURCE AND PERFORMER[a]
Calendar Years 1997–2000
(Millions of Dollars)

Source of Funds	TOTAL, All Performers	Performer				
		Federal Government	Industry	Colleges & Universities	Federally-Funded Research & Development Centers	Non-Profit Institutions
1997[r]						
All Sources—TOTAL	$212,379	$16,819	$157,539	$25,088	$5,486	$7,447
Federal Government	64,784	16,819	23,928	14,716	5,486	3,835
Gov't, Non-Federal	1,926	—	—	1,926	—	—
Industry	136,232	—	133,611	1,812	—	809
Colleges & Universities...	4,846	—	—	4,846	—	—
Nonprofit Institutions......	4,594	—	—	1,790	—	2,804
1998[r]						
All Sources—TOTAL	$226,872	$17,362	$169,180	$26,664	$5,589	$8,077
Federal Government	66,828	17,362	24,164	15,589	5,589	4,124
Gov't, Non-Federal	1,987	—	—	1,987	—	—
Industry	147,867	—	145,016	1,971	—	880
Colleges & Universities...	5,183	—	—	5,183	—	—
Nonprofit Institutions......	5,007	—	—	1,934	—	3,073
1999[r]						
All Sources—TOTAL	$244,143	$18,332	$182,823	$28,363	$5,698	$8,926
Federal Government	67,710	18,332	22,535	16,518	5,698	4,627
Gov't, Non-Federal	2,083	—	—	2,083	—	—
Industry	163,397	—	160,288	2,133	—	976
Colleges & Universities...	5,562	—	—	5,562	—	—
Nonprofit Institutions......	5,389	—	—	2,066	—	3,323
2000[p]						
All Sources—TOTAL	$264,622	$19,143	$199,855[b]	$30,154	$5,801	$9,668
Federal Government	69,626	19,143	22,210[b]	17,475	5,801	4,997
Gov't, Non-Federal	2,197	—	—	2,197	—	—
Industry	181,040	—	177,645[b]	2,310	—	1,085
Colleges & Universities...	5,969	—	—	5,969	—	—
Nonprofit Institutions......	5,789	—	—	2,203	—	3,586

Source: National Science Foundation, "Annual Survey of Industrial Research and Development" (Annually).
  a Source/performer detail not available by industry.
  b See page 104 for more current figures.

FEDERAL OUTLAYS FOR CONDUCT OF RESEARCH AND DEVELOPMENT
Fiscal Years 1989–2003
(Millions of Dollars)

Year	TOTAL	DoD	NASA	Energy[a]	Other[b]
CURRENT DOLLARS					
1989	$ 60,760	$37,819	$4,975	$5,681	$12,285
1990	63,810	38,247	6,325	5,957	13,281
1991	62,183	35,330	7,072	5,892	13,889
1992	64,728	35,504	7,617	6,043	15,564
1993	68,378	37,666	8,088	6,036	16,588
1994	68,453	35,474	7,878	5,904	19,197
1995	68,432	35,356	8,992	6,195	17,889
1996	68,439	36,936	8,083	6,135	17,285
1997	71,073	37,702	9,374	5,819	18,178
1998	72,803	37,558	9,881	5,971	19,393
1999	74,136	37,571	9,433	6,077	21,055
2000	73,947	38,279	6,369	6,282	23,017
2001	86,397	45,454	6,966	6,613	27,364
2002 [E]	97,280	50,213	6,583	8,193	32,291
2003 [E]	106,948	56,311	7,638	7,710	35,289
CONSTANT DOLLARS[c]					
1989	$ 73,293	$45,620	$6,001	$6,853	$14,819
1990	74,198	44,473	7,355	6,927	15,443
1991	69,556	39,519	7,911	6,591	15,536
1992	70,587	38,718	8,306	6,590	16,973
1993	72,820	40,113	8,613	6,428	17,666
1994	71,305	36,952	8,206	6,150	19,997
1995	69,829	36,078	9,176	6,321	18,254
1996	68,439	36,936	8,083	6,135	17,285
1997	69,748	36,999	9,199	5,711	17,839
1998	70,409	36,323	9,556	5,775	18,755
1999 [r]	70,808	35,884	9,010	5,804	20,110
2000	69,174	35,808	5,958	5,877	21,531
2001	78,973	41,548	6,367	6,045	25,013
2002 [E]	87,013	44,913	5,888	7,328	28,883
2003 [E]	93,979	49,482	6,712	6,775	31,010

Source: Office of Management and Budget, "The Budget of the United States Government" (Annually).
    a Includes defense and nondefense-related atomic energy R&D with nondefense energy R&D.
    b Includes but not limited to NSF, National Institutes of Health, DoT, & Agriculture.
    c Based on Fiscal Year GDP deflator, 1996=100.

FUNDS FOR INDUSTRIAL RESEARCH AND DEVELOPMENT
IN ALL INDUSTRIES AND THE AEROSPACE INDUSTRY
BY FUNDING SOURCE
Calendar Years 1986–2000
(Millions of Dollars)

Year	All Industries[a]			Aerospace Industry[b]		
	TOTAL	Federal Funds	Company Funds[c]	TOTAL	Federal Funds	Company Funds[c]
CURRENT DOLLARS						
1986	$ 87,823	$27,891	$ 59,932	$21,050	$14,984	$6,066
1987	92,155	30,752	61,403	24,458	18,519	5,939
1988	97,015	30,343	66,672	24,168	18,402	5,766
1989	102,055	28,554	73,501	22,331	16,828	5,503
1990	109,727	28,125	81,602	20,635	15,248	5,387
1991	116,952	26,372	90,580	16,629	11,096	5,533
1992	119,110	24,722	94,388	17,158	10,287	6,871
1993	117,400	22,809	94,591	15,056	9,372	5,684
1994	119,595	22,463	97,131	14,260	8,794	5,466
1995	132,103	23,451	108,652	16,951	11,462	5,489
1996	144,667	23,653	121,015	16,224	10,515	5,710
1997	157,539	23,928	133,611	17,865	10,904	6,961
1998	169,180	24,164	145,016	16,359	9,838	6,521
1999	182,823	22,535	160,288	14,425	9,117	5,309
2000	199,539	19,118	180,421	10,319	6,424	3,895
CONSTANT DOLLARS[d]						
1986	$116,631	$37,040	$ 79,591	$27,955	$19,899	$8,056
1987	118,756	39,629	79,128	31,518	23,865	7,653
1988	120,966	37,834	83,132	30,135	22,945	7,190
1989	122,515	34,279	88,236	26,808	20,202	6,606
1990	126,852	32,514	94,338	23,855	17,628	6,228
1991	130,381	29,400	100,981	18,538	12,370	6,168
1992	129,749	26,930	102,819	18,691	11,206	7,485
1993	124,761	24,239	100,522	16,000	9,960	6,040
1994	124,578	23,399	101,178	14,854	9,160	5,694
1995	134,662	23,905	110,756	17,279	11,684	5,595
1996	144,667	23,653	121,015	16,224	10,515	5,710
1997	154,602	23,482	131,120	17,532	10,701	6,831
1998	163,934	23,415	140,519	15,852	9,533	6,319
1999	174,616	21,523	153,093	13,777	8,708	5,071
2000	186,485	17,867	168,618	9,644	6,004	3,640

Source: National Science Foundation, "Annual Survey of Industrial Research and Development" (Annually).
  a Includes all manufacturing industries, plus those non-manufacturing industries known to conduct or finance research and development.
  b Companies classified in NAICS code 3364, having as their principal activity the manufacture of aerospace products and parts. Prior to 1999, data categorized using SIC system and reported combining codes 372 and 376.
  c Company funds include all funds for industrial R&D work performed within company facilities except funds provided by the Federal Government. Excluded are company-financed research and development contracted to outside organizations such as research institutions, universities and colleges, or other non-profit organizations.
  d Based on GDP deflator, 1996=100.

FUNDS FOR INDUSTRIAL RESEARCH AND DEVELOPMENT IN THE AEROSPACE INDUSTRY BY TYPE OF RESEARCH AND FUNDING SOURCE

Calendar Years 1964–2000
(Millions of Dollars)

Year	TOTAL	Basic Research			Applied Research			Development		
		TOTAL	Federal Funds	Com-pany Funds	TOTAL	Federal Funds	Com-pany Funds	TOTAL	Federal Funds	Com-pany Funds
1964	$ 5,078	$ 67	$ 34	$ 28	$ 766	$ 607	$ 159	$ 4,244	$ 3,948	$ 296
1965	5,148	71	41	30	735	563	172	4,342	3,921	421
1966	5,526	69	36	33	773	563	210	4,685	4,162	523
1967	5,669	71	33	38	726	490	236	4,871	4,071	800
1968	5,765	68	26	42	677	426	251	5,021	4,145	876
1969	5,882	65	24	41	597	347	250	5,220	4,216	1,004
1970	5,219	63	20	43	565	352	213	4,591	3,718	873
1971	4,881	54	37	17	461	279	182	4,365	3,583	782
1972	4,950	60	44	16	451	267	184	4,438	3,722	716
1973	5,052	50	21	29	512	308	204	4,491	3,633	858
1974	5,278	51	19	32	609	360	249	4,617	3,735	882
1975	5,713	54	17	37	614	381	233	5,044	4,119	925
1976	6,339	54	21	33	666	365	301	5,619	4,521	1,098
1977	7,033	56	25	31	753	419	334	6,223	5,017	1,206
1979 [a]	8,041	86	44	42	880	499	381	7,076	5,314	1,762
1981 [a]	11,968	131	60	71	1,484	897	587	10,353	7,738	2,615
1983	13,853	146	NA	NA	3,466	NA	NA	10,241	7,668	2,573
1984	16,033	247	NA	NA	3,067	NA	NA	12,718	9,870	2,848
1985	17,619	304	162	142	3,785	2,776	1,009	13,530	10,483	3,047
1986	21,050	311	208	103	3,198	1,571	1,627	17,541	13,205	4,336
1987	24,488	425	335	90	2,949	1,709	1,239	21,115	16,475	4,640
1988	25,900	366	263	104	2,997	1,915	1,082	22,537	17,700	4,838
1989	25,638	668	553	116	3,081	2,113	968	21,889	16,967	4,921
1990	25,356	658	519	139	3,340	1,931	1,409	21,358	16,766	4,592
1991	16,983	364	302	62	2,091	1,105	986	14,528	10,043 [b]	4,485
1992	17,158	270	235	35	1,739	976	763	15,148	9,076	6,072
1993	15,056	NA	NA	NA	1,453	825	628	NA	NA	NA
1994	14,260	NA	NA	NA	NA	NA	NA	12,787	7,978	4,809
1995	16,951	252	250	2	1,987	564	1,423	14,712	10,648	4,064
1996	16,224	NA	NA	108	NA	NA	NA	13,259	9,264	3,995
1997	17,865 [c]	NA	NA	10	NA	NA	1,508	13,275	9,115	4,159
1998	16,359 [c]	NA	NA	172	NA	NA	272	12,800	8,136	4,664
1999	14,425	NA	NA	173	NA	NA	655	11,541	7,060	4,480
2000	10,319	NA	NA	NA	NA	NA	NA	6,766	3,931	2,835

Source: National Science Foundation, "Annual Survey of Industrial Research and Development" (Annually).

a Break-outs by Research Type and Funding Source available only for odd-numbered years between 1977 and 1983.

b Computed by AIA as difference between total and company funds. Figure withheld by NSF because of imputation of more than 50 percent.

c Funding by type of research not revised nor published despite revised totals.

105

RESEARCH AND DEVELOPMENT FUNDS AS PERCENT OF NET SALES
ALL MANUFACTURING INDUSTRIES AND THE AEROSPACE INDUSTRY
Calendar Years 1978–2000

Year	All Manufacturing Industries[a]		Aerospace Industry[b]	
	Total Funds	Company Funds	Total Funds	Company Funds
1978	2.9%	2.0%	13.3%	3.2%
1979	2.6	1.9	12.9	3.5
1980	3.0	2.1	13.7	3.8
1981	3.1	2.2	16.0	4.6
1982	3.8	2.6	17.1	5.1
1983	3.9	2.6	15.2	4.1
1984	3.9	2.6	15.4	4.0
1985	4.4	3.0	14.9	3.9
1986	4.7	3.2	13.4	4.0
1987	4.6	3.1	14.7	3.6
1988	4.5	3.1	16.3	3.9
1989	4.3	3.1	13.5	3.3
1990	4.2	3.1	11.8	3.1
1991	4.2	3.2	12.1	4.0
1992	4.2	3.3	11.8	4.7
1993	3.8	3.1	12.5	4.7
1994	3.6	2.9	13.8	5.3
1995	3.6	2.9	12.9	4.2
1996	4.0	3.3	12.9	4.5
1997	3.9	3.3	8.4	3.3
1998	3.7	3.2	7.2	2.9
1999	3.7	3.2	8.8	3.2
2000	3.6	3.3	7.3	2.8

Source: National Science Foundation, "Annual Survey of Industrial Research and Development" (Annually).
a Includes all manufacturing industries known to conduct or finance research and development.
b Companies classified in NAICS code 3364, having as their principal activity the manufacture of aerospace products and parts. Prior to 1999, data categorized using SIC system and reported combining codes 372 and 376.

FEDERAL AERONAUTICS RESEARCH AND DEVELOPMENT

Fiscal Years 1985–2001
(Millions of Dollars)

Year	TOTAL	NASA[a]	DoD[b]	DoT[c]
BUDGET AUTHORITY				
1985	$ 4,335	$ 648	$3,422	$ 265
1986	6,660	601	4,927	1,132
1987	5,824	698	4,179	946
1988	6,974	723	4,989	1,262
1989	10,656	872	8,240	1,544
1990	10,690	932	7,867	1,891
1991	9,417	968	6,149	2,300
1992	11,110	1,117	7,366	2,627
1993	11,359	1,245	7,582	2,532
1994	10,703	1,546	6,848	2,309
1995	10,718	1,310	7,196	2,212
1996	10,159	1,315	6,792	2,052
1997	9,721	1,252	6,323	2,146
1998	9,682	1,327	6,256	2,099
1999	8,997	1,194	5,532	2,271
2000	9,848	1,060	6,587	2,201
2001 [E]	9,867	926	6,149	2,792
OUTLAYS				
1985	$ 4,435	$ 643	$3,101	$ 691
1986	6,073	648	4,373	1,052
1987	5,867	622	4,182	1,063
1988	6,340	679	4,448	1,213
1989	8,491	855	6,420	1,216
1990	10,009	889	7,649	1,471
1991	9,501	1,017	6,793	1,691
1992	10,011	1,122	6,790	2,099
1993	11,162	1,212	7,572	2,378
1994	11,137	1,330	7,203	2,604
1995	11,155	1,153	7,132	2,870
1996	10,837	1,187	6,974	2,676
1997	10,430	1,302	6,600	2,528
1998	10,122	1,339	6,354	2,429
1999	9,499[r]	1,217	5,913[r]	2,369
2000	9,577	1,014	6,320	2,243
2001 [E]	9,735	867	6,297	2,571

Source: NASA, "Aeronautics and Space Report of the President" (Annually).
   a   Research and Development, Construction of Facilities, Research and Program Management.
   b   Research, Development, Test, and Evaluation of aircraft and related equipment.
   c   Federal Aviation Administration: Research, Engineering, and Development; and Facilities, Engineering, and Development.

DEPARTMENT OF DEFENSE OUTLAYS
FOR RESEARCH, DEVELOPMENT, TEST, AND EVALUATION
Fiscal Years 1972–2003
(Millions of Dollars)

Year	TOTAL	Air Force	Army	Navy	Other
1972	$ 7,881	$ 3,205	$1,779	$ 2,427	$ 470
1973	8,157	3,362	1,912	2,404	479
1974	8,582	3,240	2,190	2,623	529
1975	8,866	3,308	1,964	3,021	573
1976	8,923	3,338	1,842	3,215	528
Tr.Qtr.	2,203	830	437	778	161
1977	9,795	3,618	2,069	3,481	627
1978	10,508	3,626	2,342	3,825	715
1979	11,152	4,080	2,409	3,826	837
1980	13,127	5,017	2,707	4,381	1,021
1981	15,278	6,341	2,958	4,783	1,196
1982	17,729	7,794	3,230	5,240	1,465
1983	20,554	9,182	3,658	5,854	1,861
1984	23,117	10,353	3,812	6,662	2,289
1985	27,103	11,573	3,950	8,054	3,527
1986	32,283	13,417	3,984	9,667	5,215
1987	33,596	13,347	4,721	9,176	6,352
1988	34,792	14,302	4,624	8,828	7,038
1989	37,002	14,912	4,966	9,291	7,833
1990	37,458	14,443	5,513	9,160	8,342
1991	34,589	13,050	5,559	7,586	8,371
1992	34,632	11,998	5,978	7,826	8,830
1993	36,968	12,338	6,218	8,944	9,467
1994	34,786	12,513	5,746	7,990	8,537
1995	34,710	12,052	5,081	9,230	8,347
1996	36,561	13,056	4,925	9,404	9,175
1997	37,027	14,040	4,859	8,220	9,908
1998	37,420	14,499	4,881	7,836	10,204
1999	37,363	14,172	5,027	8,052	10,112
2000	37,606	13,839	4,777	8,857	10,133
2001	40,599	14,310	5,837	9,465	10,987
2002 [E]	45,057	14,430	6,601	10,524	13,502
2003 [E]	50,823	16,460	6,837	11,784	15,742

Source: Office of Management and Budget, "The Budget of the United States Government" (Annually).

DEPARTMENT OF DEFENSE APPROPRIATIONS FOR
RESEARCH, DEVELOPMENT, TEST, AND EVALUATION
Fiscal Years 2001–2003
(Millions of Dollars)

	2001	2002^E	2003^E
TOTAL	$41,748	$48,505	$53,857

BY APPROPRIATION

	2001	2002^E	2003^E
Army	$ 6,263	$ 7,053	$ 6,918
Navy	9,596	11,389	12,502
Air Force	14,313	14,548	17,601
Defense Agencies	11,316	15,285	16,614
Director of Test & Evaluation, Defense	35	—	—
Director of Operational Test & Evaluation	225	230	222

RECAP OF BUDGET ACTIVITIES

	2001	2002^E	2003^E
Basic Research	$ 1,287	$ 1,376	$ 1,365
Applied Research	3,674	4,086	3,780
Advanced Technology Development	3,972	4,414	4,532
Demonstration and Validation	8,052	10,361	10,539
Engineering & Manufacturing Development	8,441	11,018	13,550
RDT&E Management Support	3,342	2,850	2,890
Operational Systems Development	12,980	14,399	17,200

RECAP OF FYDP PROGRAMS

	2001	2002^E	2003^E
Strategic Forces	$ 166	$ 198	$ 255
General Purpose Forces	3,042	3,278	3,701
Intelligence and Communications	8,497	9,325	11,582
Mobility Forces	345	603	675
Guard and Reserve Forces	6	13	13
Research and Development (FYDP Program 6)	28,971	34,275	36,841
Central Supply and Maintenance	341	357	254
Training Medical and Other	—	—	0
Administration and Associated Activities	96	60	101
Support of Other Nations	9	4	4
Special Operations Forces	275	392	431

Source: Department of Defense Budget, "RDT&E Programs (R-1)" (Annually).

DEPARTMENT OF DEFENSE PRIME CONTRACT AWARDS
FOR RESEARCH, DEVELOPMENT, TEST, AND EVALUATION
Fiscal Years 1997–2001
(Millions of Dollars)

Program Categories	1997	1998	1999	2000	2001
TOTAL..	$19,856	$20,103	$19,437	$19,246	$21,444
Research	1,704	1,646	1,785	1,756	2,213
Exploratory Development	1,983	2,053	2,255	2,509	2,835
Other Development / Support	16,168	16,404	15,397	14,981	16,396
Aircraft—TOTAL	$ 4,310	$ 4,609	$ 4,108	$ 4,037	$ 4,361
Research	111	207	156	261	276
Exploratory Development	127	106	110	305	319
Other Development / Support	4,072	4,297	3,842	3,471	3,766
Missile and Space Systems—TOTAL ...	4,904	5,268	4,793	4,556	5,361
Research	270	252	188	55	81
Exploratory Development	426	416	536	580	881
Other Development / Support	4,208	4,600	4,069	3,921	4,399
Electronics & Communications					
Equipment—TOTAL	3,589	2,955	3,173	2,645	3,192
Research	260	170	212	162	252
Exploratory Development	319	312	320	352	424
Other Development / Support	3,011	2,472	2,640	2,132	2,516
All Other—TOTAL[a]	7,053	7,271	7,364	8,008	8,530
Research	1,064	1,017	1,230	1,278	1,604
Exploratory Development	1,111	1,219	1,289	1,272	1,211
Other Development / Support	4,878	5,035	4,846	5,457	5,715

Source: Department of Defense, "Prime Contract Awards by Service Category and Federal Supply Classification" (Annually).
a "All Other" includes ships, tank-automotive, weapons, ammunition, services, and other.

DEPARTMENT OF DEFENSE PRIME CONTRACT AWARDS OVER $25,000 FOR RESEARCH, DEVELOPMENT, TEST, AND EVALUATION BY REGION AND TYPE OF CONTRACTOR

Fiscal Year 2001

Region	TOTAL	Type of Contractor		
		Educational Institutions	Other Non-Profit Institutions[a]	Business Firms
TOTAL (Millions of Dollars)	$21,101	$478	$1,874	$18,736
New England	$ 1,618	$ 62	$ 550	$ 1,006
Middle Atlantic	1,775	34	183	1,559
East North Central	862	56	26	781
West North Central	818	19	1	798
South Atlantic	5,490	99	630	4,756
East South Central	1,945	29	3	1,913
West South Central	579	38	55	486
Mountain	2,195	55	7	2,131
Pacific[b]	5,817	87	420	5,307
PERCENT OF TOTAL	100.0%	100.0%	100.0%	100.0%
New England	7.7%	13.0%	29.3%	5.4%
Middle Atlantic	8.4	7.1	9.7	8.3
East North Central	4.1	11.6	1.4	4.2
West North Central	3.9	4.0	0.1	4.3
South Atlantic	26.0	20.7	33.6	25.4
East South Central	9.2	6.0	0.1	10.2
West South Central	2.7	8.0	2.9	2.6
Mountain	10.4	11.5	0.4	11.4
Pacific[b]	27.6	18.1	22.4	28.3

Source: Department of Defense, Washington Headquarters Services, Directorate for Information Operations and Reports.
a Includes contracts with other government agencies.
b Includes Alaska and Hawaii.

MILITARY AIRCRAFT PROGRAMS
RESEARCH, DEVELOPMENT, TEST, AND EVALUATION[a]
BY AGENCY AND MODEL
Fiscal Years 2001, 2002, and 2003
(Millions of Dollars)

Agency and Model	2001	2002[E]	2003[E]
AIR FORCE			
B-1B	$ 148.1	$ 150.8	$ 160.7
B-2 Spirit	126.1	217.0	225.3
B-52	47.4	66.2	55.8
C-5 Galaxy	91.9	154.9	277.8
C-17 Globemaster III	168.0	109.5	157.2
C-130J Hercules	63.3	59.9	179.0
E-3 AWACS	33.2	39.0	174.0
E-8C JSTARS[b]	173.7	160.7	60.2
F-15E Eagle	91.3	107.4	81.4
F-16 Falcon	114.7	114.0	81.3
F-22 Raptor	1,411.6	881.5	627.3
*YAL-1A	386.1	475.8	598.0
ARMY			
AH-64D Longbow Apache	$ 17.0	$ 39.8	$ 46.2
*RAH-66 Comanche	590.8	781.3	910.2
UH-60 Black Hawk	28.8	71.8	99.1
DEFENSE AIRBORNE RECONNAISSANCE OFFICE			
UAVs[c]	$ 271.0	$ 519.5	$ 693.0
NAVY			
AV-8B Harrier	$ 28.1	$ 30.7	$ 18.6
CH-60S	—	78.2	113.7
E-2C Hawkeye	55.7	115.6	132.7
EA-6B Prowler	88.1	87.8	66.9
F/A-18 Hornet	221.3	251.8	204.5
H-1	133.3	170.4	241.4
*JSF[b]	682.4	1,524.9	3,471.2
MH-60R	78.4	148.1	89.0
MH-60S	30.8	44.3	23.3
V-22 Osprey	259.7	738.7	496.8

Source: Department of Defense Budget, "Program Acquisition Costs by Weapon System" (Annually) and "RDT&E Programs (R-1)" (Annually).

NOTE: See Aircraft Production Chapter for aircraft program procurement authorization data.
  a   Total Obligational Authority.
  b   Air Force and Navy funding.
  c   Air Force, Navy, and Army funding.
  *   Programs in R&D only.

EMPLOYMENT AND COST OF R&D SCIENTISTS AND ENGINEERS
ALL INDUSTRIES AND AEROSPACE INDUSTRY
Calendar Years 1979–2001

Year	Employment[a]			Cost Per R&D Scientist and Engineer[d]	
	All Industries[b] (Thousands)	Aerospace[c] (Thousands)	Aerospace as a Percent of All Industries	All Industries[b]	Aerospace[c]
1979	423.9	86.5	20.4%	$ 87,400	$ 93,300
1980	450.6	85.9	19.1	94,900	101,600
1981	487.8	95.2	19.5	103,900	128,400
1982	509.8	91.1	17.9	111,600	148,800
1983	540.9	103.1	19.1	116,000	143,600
1984	584.1	111.5	19.1	124,000	156,000
1985	622.5	130.2	20.9	130,200	161,700
1986	671.0	144.8	21.6	128,500	149,800
1987	695.8	136.3	19.6	128,800	180,400
1988	708.6	136.4	19.2	132,300	193,300
1989	722.5	134.8	18.7	134,500	207,300
1990	743.6	115.3	15.5	141,300	213,700
1991	773.4	100.2	13.0	148,600	177,000
1992	779.3	92.9	11.9	157,912	180,552
1993	764.7	97.9	12.8	153,336	176,450
1994	768.5	72.8	9.5	157,459	186,898
1995	746.1	63.5	8.5	167,339	213,328
1996	832.8	95.5	11.5	168,362	170,733
1997	885.7	94.6	10.7	171,499[r]	208,217
1998	951.5	77.0	8.1	173,589[r]	228,159[r]
1999	997.7	66.4	6.7	179,997	237,058
2000	1,033.7	55.3	5.3	192,327	256,692
2001	1,041.3	25.1	2.4	NA	NA

Source: National Science Foundation.

   a Employment as of January. Scientists and engineers working less than full time have been included in terms of their full time equivalent number.

   b All manufacturing industries and those non-manufacturing industries known to conduct or finance research and development.

   c Companies classified in NAICS code 3364, having as their principal activity the manufacture of aerospace products and parts. Prior to 1999, data categorized using SIC system and reported combining codes 372 and 376.

   d The arithmetic mean of the numbers of R&D scientists and engineers reported for January in two consecutive years, divided into the total R&D expenditures of each industry during the earlier year.

FOREIGN TRADE

Even as the manufacturing sector saw total exports and imports fall, the aerospace industry enjoyed a trade surplus in 2001. That surplus, however, has declined nearly $15 billion from the record-high level set in 1998. Slipping $700 million in 2001 to $26 billion, the aerospace trade surplus shrank for the third straight year.

Aerospace exports, at $59 billion, constituted 8% of the total value of U.S.-exported merchandise. Three years ago, when aerospace exports peaked at $64 billion, aerospace accounted for 9.4%; and ten years ago that percentage peaked at 10.4% of merchandise exports.

Civil products comprised 84% of the total value of aerospace exports. Raised by increased commercial transport deliveries, civil exports increased $3.8 billion to $49 billion. Gains in complete aircraft engine exports and aircraft and parts exports complemented the $2.5 billion more jetliner exports.

Military products, which constituted 16% of aerospace exports, increased minimally to $9.1 billion. While parts exports increased $0.5 billion, complete aircraft declined by an equal amount. In particular, fighter exports dropped by three-quarters after falling by half in 2000, while transport exports increased greatly.

Aerospace imports rose $4.5 billion to $32 billion. This is the sixth consecutive year of record imports. Foreign transport aircraft are increasingly supplying U.S. air carriers. In 1998, U.S. airlines imported just 67 civil transports. By 2001, that figure had grown to 167. In 2001, the value of imported jetliners reached $6.7 billion. This follows on a 64% increase in the previous year. Since 1997, the value of jetliner imports has increased nearly three-fold.

Similarly, imports of general aviation aircraft continued to rise. Up 26% in 2001, general aviation imports grew to $6.3 billion.

General aviation aircraft imports, in dollars, have more than doubled since 1997.

The United Kingdom was the largest importer of U.S. aerospace exports—$6.5 billion in 2001. The next five largest in order were: France, $5.2 billion; Germany, $4.4 billion; Singapore, $4.2 billion; Canada, $4.1 billion; and Japan, $3.8 billion. Five of these six countries, which imported a total of $28 billion of U.S. aerospace exports, are also leading aerospace producers. The U.S. imported from France, $8.7 billion; Canada, $8.0 billion; United Kingdom, $4.8 billion; Germany, $3.8 billion; and Japan, $2.0 billion. Their combined aerospace exports to the United States totaled $27 billion indicating nearly balanced trade between aerospace-producing countries.

The Export-Import Bank of the United States (Eximbank) promotes U.S. exports by offering foreign customers loans and loan guarantees competitive with the official export credit organizations of our foreign competitors. In FY 2001, Eximbank loans and guarantees enabled the purchase of 60 jet aircraft worth $3.1 billion.

U.S. TOTAL AND AEROSPACE FOREIGN TRADE[a]
Calendar Years 1967–2001
(Millions of Dollars)

Year	Total U.S. Merchandise Trade			Aerospace		
	Trade Balance	Exports	Imports	Trade Balance	Exports	Imports
1967	$ 4,122	$ 30,934	$ 26,812	$ 1,961	$ 2,248	$ 287
1968	837	34,063	33,226	2,661	2,994	333
1969	1,289	37,332	36,043	2,831	3,138	307
1970	3,225	43,176	39,952	3,097	3,405	308
1971	(1,476)[b]	44,087	45,563	3,830	4,203	373
1972	(5,729)	49,854	55,583	3,230	3,795	565
1973	2,390	71,865	69,476	4,360	5,142	782
1974	(3,884)	99,437	103,321	6,350	7,095	745
1975	9,551	108,856	99,305	7,045	7,792	747
1976	(7,820)	116,794	124,614	7,267	7,843	576
1977	(28,353)	123,182	151,534	6,850	7,581	731
1978	(30,205)	145,847	176,052	9,058	10,001	943
1979	(23,922)	186,363	210,285	10,123	11,747	1,624
1980	(19,696)	225,566	245,262	11,952	15,506	3,554
1981	(22,267)	238,715	260,982	13,134	17,634	4,500
1982	(27,510)	216,442	243,952	11,035	15,603	4,568
1983	(52,409)	205,639	258,048	12,619	16,065	3,446
1984	(106,703)	223,976	330,678	10,082	15,008	4,926
1985	(117,712)	218,815	336,526	12,593	18,725	6,132
1986	(138,279)	227,159	365,438	11,826	19,728	7,902
1987	(152,119)	254,122	406,241	14,575	22,480	7,905
1988	(118,526)	322,426	440,952	17,860	26,947	9,087
1989	(109,399)	363,812	473,211	22,083	32,111	10,028
1990	(101,718)	393,592	495,311	27,282	39,083	11,801
1991	(66,723)	421,730	488,453	30,785	43,788	13,003
1992	(84,501)	448,164	532,665	31,356	45,018	13,662
1993	(115,568)	465,091	580,659	27,235	39,418	12,183
1994	(150,630)	512,626	663,256	25,010	37,373	12,363
1995	(158,801)	584,742	743,543	21,561	33,071	11,509
1996	(170,214)	625,075	795,289	26,602	40,270	13,668
1997	(180,522)	689,182	869,704	32,239	50,374	18,134
1998	(229,758)	682,138	911,896	40,960	64,071	23,110
1999	(328,821)	695,797	1,024,618	37,381	62,444	25,063
2000	(436,104)	781,918	1,218,022	26,734	54,679	27,944
2001	(411,042)	730,912	1,141,954	26,035	58,508	32,473

Source: Bureau of the Census, Foreign Trade Division and Aerospace Industries Association, based on data from International Trade Administration.

NOTE: The Commerce Department began reporting international trade using the Harmonized Tariff Schedules of the United States in 1989. Previous years based on the Tariff Schedules of the United States Annotated.

a Total U.S. and aerospace foreign trade are reported as (1) exports of domestic merchandise, including Department of Defense shipments and undocumented exports to Canada, free alongside-ship basis, (2) imports for consumption, customs value basis.

b First U.S. trade deficit since 1888.

TOTAL U.S. EXPORTS AND EXPORTS OF AEROSPACE PRODUCTS
Calendar Years 1967–2001
(Millions of Dollars)

Year	Total Exports of U.S. Merchandise[a]	Exports of Aerospace Products				
		TOTAL	Percent of Total U.S. Exports	Civil		Military
				Total	Transports	
1967	$ 30,934	$ 2,248	7.3%	$ 1,380	$ 611	$ 868
1968	34,063	2,994	8.8	2,289	1,200	705
1969	37,332	3,138	8.4	2,027	947	1,111
1970	43,176	3,405	7.9	2,516	1,283	889
1971	44,087	4,203	9.5	3,080	1,567	1,123
1972	49,854	3,795	7.6	2,954	1,119	841
1973	71,865	5,142	7.2	3,788	1,664	1,354
1974	99,437	7,095	7.1	5,273	2,655	1,822
1975	108,856	7,792	7.2	5,324	2,397	2,468
1976	116,794	7,843	6.7	5,677	2,468	2,166
1977	123,182	7,581	6.2	5,049	1,936	2,532
1978	145,847	10,001	6.9	6,018	2,558	3,983
1979	186,363	11,747	6.3	9,772	4,998	1,975
1980	225,566	15,506	6.9	13,248	6,727	2,258
1981	238,715	17,634	7.4	13,312	7,180	4,322
1982	216,442	15,603	7.2	9,608	3,834	5,995
1983	205,639	16,065	7.8	10,595	4,683	5,470
1984	223,976	15,008	6.7	9,659	3,195	5,350
1985	218,815	18,725	8.6	12,942	5,518	5,783
1986	227,159	19,728	8.7	14,851	6,276	4,875
1987	254,122	22,480	8.8	15,768	6,377	6,714
1988	322,426	26,947	8.4	20,298	8,766	6,651
1989	363,812	32,111	8.8	25,619	12,313	6,492
1990	393,592	39,083	9.9	31,517	16,691	7,566
1991	421,730	43,788	10.4	35,548	20,881	8,239
1992	448,164	45,018	10.0	36,906	22,379	8,111
1993	465,091	39,418	8.5	31,823	18,146	7,596
1994	512,626	37,373	7.3	30,050	15,931	7,322
1995	584,742	33,071	5.7	25,079	10,606	7,991
1996	625,075	40,270	6.4	29,477	13,624	10,792
1997	689,182	50,374	7.3	40,075	21,028	10,299
1998	682,138	64,071	9.4	51,999	29,168	12,072
1999	695,797	62,444	9.0	50,624	25,694	11,820
2000	781,918	54,679	7.0	45,566	19,615	9,113
2001	730,912	58,508	8.0	49,371	22,151	9,137

Source: Bureau of the Census, Foreign Trade Division and Aerospace Industries Association, based on data from International Trade Administration.

NOTE: International trade reported using Harmonized Tariff Schedules after 1988.

a Includes DoD shipments and undocumented exports to Canada, free alongside-ship basis.

U.S. EXPORTS OF AEROSPACE PRODUCTS[a]
BY MAJOR COUNTRIES OF DESTINATION
Calendar Years 1997–2001
(Millions of Dollars)

Country of Destination	1997	1998	1999	2000	2001
Australia	$ 885	$1,050	$1,426	$1,284	$1,112
Brazil	1,045	1,461	1,575	1,636	2,473
Canada	2,796	3,107	3,438	3,747	4,071
China	2,256	3,731	2,491	1,794	2,591
France	2,688	4,286	5,322	4,691	5,248
Germany	2,519	4,214	4,325	4,581	4,417
Israel	716	1,595	1,789	877	1,275
Italy	629	587	1,426	873	823
Japan	5,071	6,057	5,401	4,257	3,795
Korea, South	2,479	1,888	1,899	2,157	2,844
Malaysia	1,440	1,382	539	353	696
Netherlands	1,468	1,037	1,566	1,796	1,285
Saudi Arabia	2,625	5,008	3,299	1,960	1,432
Singapore	2,030	2,296	2,069	1,387	4,160
Spain	364	281	1,305	1,381	720
Sweden	443	792	1,295	1,347	608
Taiwan	2,407	2,915	2,237	1,622	1,609
Thailand	1,186	824	618	572	575
Turkey	541	883	957	1,307	842
United Kingdom	6,471	7,569	7,845	6,478	6,536

Source: Aerospace Industries Association, based on data from the International Trade Administration.
a Includes all civil products, free alongside-ship basis; excludes military products whose country of destination are not reported.

U.S. IMPORTS OF AEROSPACE PRODUCTS[a]
BY MAJOR COUNTRIES OF ORIGIN
Calendar Years 1997–2001
(Millions of Dollars)

Country of Origin	1997	1998	1999	2000	2001
Brazil	$ 371	$ 917	$1,285	$1,494	$1,973
Canada	3,800	4,867	5,087	6,253	7,985
France	4,087	5,814	6,313	8,071	8,721
Germany	1,187	2,044	2,707	3,364	3,775
Israel	439	493	428	515	576
Italy	480	643	736	506	527
Japan	1,728	2,148	1,710	1,614	1,986
Korea, South	135	196	186	195	302
Netherlands	227	225	161	173	173
Singapore	276	325	87	96	122
Sweden	287	306	147	132	159
United Kingdom	4,034	5,173	4,968	4,197	4,818

Source: Aerospace Industries Association, based on data from the International Trade Administration.
   a   Includes civil and military products, c.i.f. (Cost, Insurance, and Freight) basis.

U.S. EXPORTS OF AEROSPACE PRODUCTS
Calendar Years 1998–2001
(Millions of Dollars)

Use and Type	1998	1999	2000	2001
TOTAL..................................	$64,071	$62,444	$54,679	$58,508
CIVIL—TOTAL	$51,999	$50,624	$45,566	$49,371
Complete Aircraft—TOTAL	$31,427	$28,450	$22,156	$24,787
Transports............................	29,168	25,694	19,615	22,151
General Aviation[a]	813	1,309	1,136	1,357
Helicopters	148	137	170	170
Used Aircraft	1,270	1,286	1,208	1,078
Other, incl. Spacecraft[b]	698	435	167	188
Aircraft Engines—TOTAL	3,158	3,714	4,610	5,258
Turbine Engines	3,071	3,602	4,510	5,142
Piston Engines	87	112	101	116
Aircraft and Engine Parts incl. Spares—TOTAL	16,744	18,051	18,660	19,169
Aircraft Parts & Accessories	10,840	11,943	12,289	12,606
Aircraft Engine Parts	5,904	6,108	6,371	6,563
MILITARY—TOTAL	$12,072	$11,820	$ 9,113	$ 9,137
Complete Aircraft—TOTAL[c]	$ 3,821	$ 4,221	$ 2,556	$ 2,096
Fighters & Fighter Bombers	2,514	2,543	1,287	339
Transports............................	618	878	408	645
Helicopters	358[r]	358	594	572
Used Aircraft	213	303	85	247
Other, incl. Spacecraft[b]	697	509	303	432
Aircraft Engines—TOTAL	367	581	333	281
Turbine Engines	256	500	248	147
Piston Engines	111	81	85	134
Aircraft and Engine Parts incl. Spares—TOTAL	6,382	5,557	5,049	5,503
Aircraft Parts & Accessories	5,311	4,558	4,151	4,602
Aircraft Engine Parts	1,071	999	898	901
Guided Missiles, Rockets, & Parts—TOTAL	923	1,091	1,053	1,119
Guided Missiles & Rockets	491	576	402	223
Missile & Rocket Parts	431	505	643	893
Missile & Rocket Engines	2	10	9	3
Missile & Rocket Engine Parts ...	—	—	—	—

Source: Aerospace Industries Association, based on data from International Trade Administration.
   a All fixed-wing aircraft under 33,000 pounds.
   b Products within this category are not designated civil or military by the Harmonized Tariff Schedules. Historically, aircraft herein have been predominantly civil. Also, spacecraft not included in "Complete Aircraft—Total."
   c Includes aircraft exported under Military Assistance Programs and Foreign Military Sales.

U.S. IMPORTS OF AEROSPACE PRODUCTS
Calendar Years 1998–2001
(Millions of Dollars)

Use and Type	1998	1999	2000	2001
TOTAL	$23,110	$25,063	$27,944	$32,473
CIVIL—TOTAL	$16,837	$18,709	$21,994	$25,670
Complete Aircraft—TOTAL	$ 6,933	$ 8,773	$12,388	$14,709
Transports................................	2,405	3,397	5,560	6,686
General Aviation........................	3,530	4,279	5,005	6,283
Helicopters	536	432	489	419
Other, including Used Aircraft, & Gliders, Balloons, & Airships[a] ...	461	665	1,334	1,321
Aircraft Engines—TOTAL	2,039	2,257	1,864	2,418
Turbine Engines[b]........................	2,006	2,233	1,846	2,394
Piston Engines	33	23	17	23
Aircraft & Engine Parts—TOTAL ...	7,866	7,680	7,742	8,543
Aircraft Parts and Accessories[b]	4,901	4,848	4,679	5,250
Turbine Engine Parts[b].................	2,688	2,327	2,554	2,880
Piston Engine Parts	130	110	121	147
Spacecraft, Other Parts & Accessories[bc]	147	396	389	266
MILITARY—TOTAL	$ 6,273	$ 6,354	$ 5,951	$ 6,804
Complete Aircraft—TOTAL	$ 6	$ 7	$ 11	$ 2
Aircraft Engines—TOTAL	2,037	2,257	1,884	2,432
Turbine Engines[b]........................	2,006	2,233	1,846	2,394
Piston Engines including Parts	31	24	38	37
Aircraft & Engine Parts—TOTAL ...	4,230	4,091	4,056	4,370
Aircraft Parts[b]	1,252	1,311	1,129	1,216
Turbine Engine Parts[b].................	2,737	2,361	2,543	2,856
Spacecraft, Missiles, Rockets, Other Parts, & Accessories[bc]	240	418	384	298

Source: Aerospace Industries Association, based on data from International Trade Administration.
  a Products within this category are not designated civil or military by the Harmonized Tariff Schedules. Historically, these products have been predominantly civil.
  b Category contains products whose use (civil or military) is unspecified by the Harmonized Tariff Schedules. Figures for those products distributed equally between civil and military.
  c Includes satellites, propulsion engines, and parts.

U.S. EXPORTS OF MILITARY AIRCRAFT[a]
Calendar Years 1997–2001

Type of Aircraft	1997	1998	1999	2000	2001
NUMBER OF AIRCRAFT	396	364	309	344	260
Fighters and Fighter Bombers	45	65	68	36	13
Transports	—	12	17	8	3
Helicopters	71	29	75	63	37
New Aircraft, NEC........................	221	163	66	88	96
Used or Rebuilt Aircraft	59	95	83	149	111
VALUE (Millions of Dollars)............	$2,397	$3,821	$4,221	$2,556	$2,096
Fighters and Fighter Bombers	$1,823	$2,514	$2,543	$1,287	$ 339
Transports	—	618	878	408	645
Helicopters	391	360	358	594	572
New Aircraft, NEC........................	49	119	140	181	293
Used or Rebuilt Aircraft	133	213	303	85	247

Source: Aerospace Industries Association, based on data from the International Trade Administration.
   a   Includes aircraft exported under Military Assistance Programs and Foreign Military Sales.
 NEC   Not elsewhere classified.

U.S. EXPORTS OF CIVIL AIRCRAFT
Calendar Years 1997–2001

Type and Size	1997	1998	1999	2000	2001
NUMBER OF AIRCRAFT[a]	1,431	1,518	1,451	1,472	2,047
Helicopters—TOTAL	259	238	181	304	309
Under 2,200 lbs	199	196	147	259	277
Over 2,200 lbs	60	42	34	45	32
General Aviation—TOTAL	409	399	503	411	446
Single-Engine	188	208	253	186	194
Multi-Engine, under 4,400 lbs	35	64	66	61	70
Multi-Engine, 4,400-10,000 lbs	102	48	113	60	52
Multi-Engine, 10,000-33,000 lbs	84	79	71	104	130
Transports—TOTAL	252	375	341	268	253
Passenger Aircraft, over 33,000 lbs	239	362	326	255	238
Cargo Aircraft, over 33,000 lbs	10	13	13	12	15
Other, over 33,000 lbs, incl. Pass./Cargo Combi	3	—	2	1	—
Other Aircraft—TOTAL[a]	511	506	426	489	472
Used or Rebuilt Aircraft	511	506	426	489	472
Other Aircraft, including Balloons, Gliders, & Kites	452	526	563	535	567
VALUE (Millions of Dollars)	$23,112	$31,427	$28,450	$22,156	$24,787
Helicopters—TOTAL	$ 207	$ 148	$ 137	$ 170	$ 170
Under 2,200 lbs	32	47	24	70	54
Over 2,200 lbs	175	101	113	120	116
General Aviation—TOTAL	946	813	1,309	1,136	1,357
Single-Engine	90	100	140	80	102
Multi-Engine, under 4,400 lbs	14	21	24	21	39
Multi-Engine, 4,400-10,000 lbs	349	206	519	227	199
Multi-Engine, 10,000-33,000 lbs	493	486	627	809	1,016
Transports—TOTAL	21,028	29,168	25,694	19,615	22,151
Passenger Aircraft, over 33,000 lbs	19,266	27,700	23,733	17,472	19,487
Cargo Aircraft, over 33,000 lbs	1,251	1,468	1,621	1,963	2,664
Other, over 33,000 lbs, incl. Pass./Cargo Combi	512	—	340	180	—
Other Aircraft—TOTAL	932	1,298	1,311	1,235	1,109
Used or Rebuilt Aircraft	909	1,270	1,286	1,208	1,078
Other Aircraft, including Balloons, Gliders, & Kites	22	28	25	27	31

Source: Aerospace Industries Association, based on data from International Trade Administration.
 a Numbers of gliders, balloons, & kites excluded from civil aircraft totals.

U.S. IMPORTS OF COMPLETE AIRCRAFT
Calendar Years 1998–2001

Use and Type	1998	1999	2000	2001
NUMBER OF AIRCRAFT	2,024	1,928	2,159	2,140
Civil Aircraft—TOTAL	1,997	1,893	2,143	2,129
New Complete Aircraft:				
Helicopters	274	217	238	229
General Aviation:				
Single-Engine	102	162	142	144
Multi-Engine, under 4,400 lbs	4	3	4	—
Multi-Engine, 4,400-10,000 lbs	3	1	7	14
Multi-Engine, Turbojet/Turbofan, 10,000-33,000 lbs	171	239	286	345
Multi-Engine, Other, including Turboshaft, 10,000-33,000 lbs ...	60	27	21	16
Transports, Multi-Engine, over 33,000 lbs................................	67	98	147	167
Other Civil Aircraft:				
Gliders[a]	169	92	133	122
Balloons & Airships[a]	170	101	75	80
Others including Kites[a]	666	683	849	739
Used or Rebuilt	311	270	241	273
Military Aircraft—TOTAL	27	35	16	11
New Complete Aircraft	7	9	2	—
Used or Rebuilt	20	26	14	11
TOTAL VALUE (Millions of Dollars) ...	$6,939.0	$8,779.6	$12,399.1	$14,711.1
Civil Aircraft—TOTAL	$6,932.7	$8,772.7	$12,388.3	$14,709.0
New Complete Aircraft:				
Helicopters	535.7	431.7	489.3	419.1
General Aviation:				
Single-Engine	81.3	145.5	134.8	161.2
Multi-Engine, under 4,400 lbs	3.6	0.2	2.5	—
Multi-Engine, 4,400-10,000 lbs	6.4	4.5	15.0	35.8
Multi-Engine, Turbojet/Turbofan, 10,000-33,000 lbs	2,860.8	3,879.1	4,647.8	5,879.4
Multi-Engine, Other, including Turboshaft, 10,000-33,000 lbs ...	578.3	249.7	205.0	206.4
Transports, Multi-Engine, over 33,000 lbs	2,405.4	3,396.8	5,559.6	6,685.8
Other Civil Aircraft:				
Gliders[a]	2.3	1.6	2.3	2.0
Balloons & Airships[a]	12.2	7.2	8.2	9.9
Others including Kites[a]	3.6	7.0	9.5	9.0
Used or Rebuilt	443.1	649.4	1,314.4	1,300.4
Military Aircraft—TOTAL	$ 6.3	$ 6.9	$ 10.9	$ 2.1
New Complete Aircraft	1.0	2.5	0.2	—
Used or Rebuilt	5.3	4.4	10.7	2.1

Source: Aerospace Industries Association, based on data from International Trade Administration.
   a Products within this category are not designated civil or military by the Harmonized Tariff Schedules. Historically, these products have been predominantly civil.

U.S. EXPORTS OF COMMERCIAL TRANSPORT AIRCRAFT[a]
Calendar Years 1997–2001

Region of Destination	1997	1998	1999	2000	2001
NUMBER OF AIRCRAFT	252	375	341	268	253
Canada & Greenland	—	—	4	—	9
Latin America & Caribbean	11	23	19	12	17
Europe	91	150	169	142	85
Middle East	18	36	55	38	28
Asia	123	150	81	57	81
Oceania	5	9	9	11	12
Africa	4	7	4	8	21
VALUE (Millions of Dollars)	$21,028	$29,168	$25,694	$19,615	$22,151
Canada & Greenland	$ —	$ —	$ 237	$ —	$ 510
Latin America & Caribbean	505	1,215	807	607	1,184
Europe	7,538	9,885	11,852	10,015	6,892
Middle East	2,449	3,871	3,797	2,231	2,570
Asia	9,916	12,894	7,872	5,618	8,818
Oceania	473	628	725	710	816
Africa	147	674	404	433	1,360

Source: Aerospace Industries Association, based on data from the International Trade Administration.
   a   Airframe weight exceeding 33,000 pounds.

U.S. EXPORTS OF CIVIL HELICOPTERS[a]
Calendar Years 1997–2001

Region of Destination	1997	1998	1999	2000	2001
NUMBER OF AIRCRAFT	259	238	181	304	309
Canada & Greenland	9	6	9	14	6
Latin America & Caribbean	36	57	25	55	59
Europe	100	133	100	126	123
Middle East	2	—	1	4	16
Asia	61	26	25	60	46
Oceania	48	14	17	23	49
Africa	3	2	4	22	10
VALUE (Millions of Dollars)	$207.1	$148.1	$136.6	$169.9	$169.7
Canada & Greenland	$ 4.4	$ 8.4	$ 4.8	$ 6.1	$ 2.0
Latin America & Caribbean	21.9	25.6	15.2	24.7	12.2
Europe	56.5	65.8	56.4	60.8	53.7
Middle East	1.1	—	1.6	13.9	13.9
Asia	116.3	43.4	55.1	56.9	72.5
Oceania	4.6	4.7	1.4	3.1	6.6
Africa	2.4	0.3	2.0	4.4	8.8

Source: Aerospace Industries Association, based on data from the International Trade Administration.
   a   Excludes used helicopters.

U.S. IMPORTS OF CIVIL HELICOPTERS[a]
Calendar Years 1997–2001

Country of Origin	1997	1998	1999	2000	2001
NUMBER OF AIRCRAFT	240	274	217	238	229
Canada	204	189	146	144	113
France	26	56	47	64	80
Germany	9	12	8	14	11
Italy	1	11	13	12	23
Others[b]	—	6	3	4	2
VALUE (Millions of Dollars).........	$460.1	$535.7	$431.7	$489.3	$419.1
Canada	$415.3	$419.1	$330.0	$355.0	$267.4
France	23.7	60.4	50.0	60.1	70.4
Germany	18.3	27.8	20.3	38.0	29.2
Italy	2.9	27.4	31.4	35.3	50.8
Others[b]	—	1.0	0.1	0.9	1.4

Source: Aerospace Industries Association, based on data from the International Trade Administration.
  a Excludes used helicopters.
  b Includes 2 from Australia, 2 from Japan, 1 from Papua New Guinea, and 1 from Poland in 1998; 2 from Japan and 1 from New Zealand in 1999; 3 from Japan and 1 from Switzerland in 2000; and 2 from Sweden in 2001.

U.S. EXPORTS OF GENERAL AVIATION AIRCRAFT[a]
Calendar Years 1997–2001

Region of Destination	1997	1998	1999	2000	2001
NUMBER OF AIRCRAFT	409	399	503	411	446
Canada & Greenland	31	25	22	39	33
Latin America & Caribbean	117	117	181	89	100
Europe	131	140	189	180	191
Middle East	1	10	19	18	30
Asia	44	35	26	23	41
Oceania	45	36	32	26	18
Africa	40	36	34	36	33
VALUE (Millions of Dollars).........	$945.9	$813.0	$1,309.0	$1,136.4	$1,357.0
Canada & Greenland	$116.0	$101.4	$ 54.6	$ 195.1	$ 141.4
Latin America & Caribbean	282.0	192.8	324.4	234.6	252.9
Europe	220.9	256.0	571.3	449.4	612.9
Middle East	10.8	11.9	96.6	67.8	78.0
Asia	156.5	137.8	97.8	103.2	170.2
Oceania	74.4	52.3	55.7	15.6	30.5
Africa	85.3	60.8	108.5	70.7	71.3

Source: Aerospace Industries Association, based on data from the International Trade Administration.
   a   All fixed-wing aircraft under 33,000 pounds.

U.S. IMPORTS OF GENERAL AVIATION AIRCRAFT[a]
Calendar Years 1997–2001

Country of Origin	1997	1998	1999	2000	2001
NUMBER OF AIRCRAFT	282	340	432	460	519
Brazil	21	58	84	92	117
Canada	87	104	139	134	153
France	50	57	78	84	106
Germany	38	27	26	39	39
Israel	5	9	19	32	33
Poland	10	13	9	2	—
Russia	4	4	3	1	1
Sweden	19	20	1	—	—
Switzerland	25	30	50	56	56
United Kingdom....................	14	1	—	—	1
Other	9	17	23	20	13
VALUE (Millions of Dollars)......	$2,513.7	$3,530.4	$4,279.1	$5,005.0	$6,282.7
Brazil	$ 256.5	$ 782.6	$1,162.8	$1,411.5	$1,877.4
Canada	1,155.2	1,521.9	1,737.5	1,839.8	2,603.3
France	748.6	857.3	1,015.8	1,201.4	1,160.3
Germany	26.1	39.6	97.6	191.3	209.0
Israel	40.0	81.8	135.1	234.6	283.8
Poland	1.3	1.9	1.3	0.2	—
Russia	0.2	0.4	0.4	0.1	0.1
Sweden	153.5	176.6	9.0	—	—
Switzerland	57.0	66.8	112.7	119.6	128.3
United Kingdom....................	74.8	0.1	—	—	0.0
Other	0.6	1.3	6.9	6.4	20.4

Source: Aerospace Industries Association, based on data from the International Trade Administration.
   a   All fixed-wing aircraft under 33,000 pounds.

U.S. EXPORTS OF AIRCRAFT ENGINES
Calendar Years 1999–2001
(Values in Millions of Dollars)

Type of Engine	1999		2000		2001	
	Number	Value	Number	Value	Number	Value
TOTAL..............................	19,989	$4,295	21,501	$4,943	31,495	$5,539
Turbine Engines—TOTAL......	7,980	$4,103	10,424	$4,758	9,962	$5,289
Civil	5,270	3,602	8,958	4,510	8,717	5,142
Military...........................	2,710	500	1,466	248	1,245	147
Piston Engines—TOTAL	12,009	193	11,077	185	21,533	250
Civil, New, Under 500 HP	842	19	705	14	1,894	20
Civil, New, Over 500 HP ...	239	7	531	10	830	18
Civil, Used	3,941	85	3,660	77	8,568	78
Military...........................	6,987	81	6,181	85	10,241	34

Source: Aerospace Industries Association, based on data from the International Trade Administration.

U.S. IMPORTS OF AIRCRAFT ENGINES[a]
Calendar Years 1999–2001
(Values in Millions of Dollars)

Type of Engine	1999		2000		2001	
	Number	Value	Number	Value	Number	Value
TOTAL	7,726	$4,497	10,412	$3,726	7,999	$4,826
Turbine Engines	3,565	$4,465	4,459	$3,692	3,375	$4,789
Piston Engines—TOTAL	4,161	32	5,953	34	4,624	38
Military	2,252	9	2,736	17	1,756	14
Civil, New, Small	336	2	892	2	1,342	4
Civil, New, Large	824	3	1,691	4	948	4
Civil, Used	749	18	634	11	578	15

Source: Aerospace Industries Association, based on data from the International Trade Administration.
   a   New and used.

EXPORT-IMPORT BANK
TOTAL AUTHORIZATIONS OF LOANS AND GUARANTEES
AND AUTHORIZATIONS IN SUPPORT OF AIRCRAFT EXPORTS
Fiscal Years 1987–2001
(Millions of Dollars)

Year	Total Authori-zations	Authorizations in Support of Aircraft Exports			
		TOTAL	Percent of Total Authori-zations	Commercial Jet Aircraft[a]	Other Aircraft[b]
LOANS[c]					
1987	$ 599	$ 17.0	2.8%	$ 13.3	$ 3.7
1988	685	—	—	—	—
1989	695	166.4	23.9	158.0	8.4
1990	614	5.0	0.8	—	5.0
1991	777	—	—	—	—
1992	817	—	—	—	—
1993	1,748	—	—	—	—
1994	3,016	—	—	—	—
1995	1,598	—	—	—	—
1996	1,236	—	—	—	—
1997	1,549	—	—	—	—
1998	103	—	—	—	—
1999	903	590.8	65.4	590.8	—
2000	933	75.7	8.1	75.7	—
2001	873	—	—	—	—
GUARANTEES[d]					
1987	$1,514	$ 808.3	53.4%	$ 808.3	$ —
1988	601	89.2	14.8	73.4	15.8
1989	1,292	496.4	38.4	390.4	106.0
1990	3,333	1,666.3	50.0	224.7	1,441.6
1991	6,034	606.0	10.1	566.9	40.0
1992	7,301	1,667.0	22.8	1,597.1	69.9
1993	9,095	3,488.6	38.4	3,488.6	—
1994	7,609	2,959.0	38.9	2,959.0	—
1995	5,712	977.0	17.1	977.0	—
1996	6,412	1,155.0	18.0	1,155.0	—
1997	7,761	1,959.0	25.2	1,959.0	—
1998	6,151	2,542.5	41.3	2,542.5	—
1999	8,299	5,543.8	66.8	5,543.8	—
2000	8,413	3,647.4	43.4	3,437.8	209.6
2001	6,101	2,736.5	44.8	2,540.5	196.0

Source: Export-Import Bank of the United States.
  a Includes complete aircraft, engines, parts, and retrofits.
  b Includes business aircraft, general aviation aircraft, helicopters, and related goods and services.
  c Loans are commitments for direct financing by the Export-Import Bank to foreign buyers of U.S. equipment and services, which are made to commercial banks and may subsequently be guaranteed by the Export-Import Bank, in which case the value of the loans is also included with Guarantees.
  d Guarantees by the Export-Import Bank provide assurances of repayment of principal and interest on loans made by private lending institutions, such as commercial banks, for major export transactions. Excludes insurance.

EXPORT-IMPORT BANK
SUMMARY OF COMMERCIAL JET AIRCRAFT AUTHORIZATIONS
FOR LOANS[a] AND GUARANTEES[b]
Fiscal Years 1976–2001
(Values in Millions of Dollars)

Year	No. of Jet Aircraft[c]		Export Value[c]		No. of New Commitments		Gross Authorizations	
	Loans	Guar-antees	Loans	Guar-antees	Loans	Guar-antees	Loans	Guar-antees
New Authorizations:								
1976	77	6	$1,017	$ 139	34	11	$ 398	$ 87
Tr.Qtr.	15	5	219	182	6	3	94	59
1977	31	25	330	902	16	14	138	294
1978	29	5	479	253	18	5	189	77
1979	118	7	2,938	317	35	10	1,399	239
1980	136	21	3,975	901	36	24	1,693	1,088
1981	121	18	4,568	637	26	17	2,550	533
1982	11	6	441	113	5	2	199	78
1983	21	9	779	619	3	4	384	601
1984	37	8	1,023	327	7	4	532	294
1985	—	14	19	481	1	5	13	289
1986	3	13	74	451	1	9	46	277
1987	—	27	22	1,449	1	14	13	808
1988	—	2	—	94	—	2	—	73
1989	3	5	253	459	1	2	158	390
1990	—	6	—	264	—	2	—	225
1991	—	12	—	665	—	3	—	567
1992	—	37	—	1,889	—	12	—	1,597
1993	—	70	—	4,122	—	27	—	3,489
1994	—	59	—	3,507	—	19	—	2,959
1995	—	27	—	1,205	—	12	—	974
1996	—	18	—	1,089	—	8	—	923
1997	—	34	—	2,357	—	14	—	1,959
1998	—	65	—	3,059	—	24	—	2,543
1999	17	106	1,170	6,464	2	32	591	5,544
2000	5	53	150	4,047	2	17	76	3,438[r]
2001	—	60	—	3,052	—	12	—	2,540

Source: Export-Import Bank of the United States.

a Loans are commitments for direct financing by the Export-Import Bank to foreign buyers of U.S. equipment and services, which are made by the Export-Import Bank to commercial banks and which subsequently may be guaranteed by the Export-Import Bank in which case the value of the loans is included with Guarantees.

b Guarantees by the Export-Import Bank provide assurances of repayment of principal and interest on loans made by private lending institutions, such as commercial banks, for major export transactions. Excludes insurance.

c For Export-Import Bank commitments including both loan and guarantee authorization, number of aircraft and export value reported under "Loans."

EXPORT-IMPORT BANK
AUTHORIZATIONS OF LOANS AND GUARANTEES
IN SUPPORT OF EXPORTS OF COMMERCIAL JET AIRCRAFT
Fiscal Years 2000–2001
(Values in Millions of Dollars)

| Customer (Country/Airline) | Number and Aircraft Model or Related Product | Export Value | Loans (Direct Credits) | | | | Guar- antees |
			Amount	Percent Cover- age[a]	Interest Rate	Repay- ment Terms[b]	Amount
FY 2001							
TOTAL	60 aircraft	$3,052	—	—	—	—	$2,540
Algeria/Air Algerie	7 x 737	$ 224	—	—	—	—	$ 191
Austria/Austrian Airlines ...	2 x 737, 1 x 767	166	—	—	—	—	137
Chile/Lan Chile	2 x 767	179	—	—	—	—	153
China/Hainan Airlines	5 x 737	290	—	—	—	—	247
China/Shanghai Airlines ...	1 x 767	80	—	—	—	—	67
India/Jet Airways	10 x 737	405	—	—	—	—	346
Korea/Korean Air Lines ...	4 x 747, 1 x 777	795	—	—	—	—	655
Romania/Tarom Romanian Air Transport	2 x 737	90	—	—	—	—	70
South Africa/South African Airways	16 x 737	320	—	—	—	—	258
Thailand/Thai Airways International	2 x 777	255	—	—	—	—	208
Turkey/Turk Hava Yollari Tao	2 x 737	82	—	—	—	—	68
FY 2000							
TOTAL	58 aircraft	$4,194	$76	1.8%	NA	NA	$3,438
Algeria/Air Algerie	5 x 737	$ 129	$66	50.8%	7.30%	16-S	$ 66
Azerbaijan/Azerbaijan Airlines	2 x 757	68	—	—	—	—	66
Chile/Ladeco	1 x 767	84	—	—	—	—	72

(Continued on next page)

EXPORT-IMPORT BANK
AUTHORIZATIONS OF LOANS AND GUARANTEES
IN SUPPORT OF EXPORTS OF COMMERCIAL JET AIRCRAFT
Fiscal Years 2000–2001, continued
(Values in Millions of Dollars)

| Customer (Country/Airline) | Number and Aircraft Model or Related Product | Export Value | Authorizations | | | | Guar- antees |
| | | | Loans (Direct Credits) | | | | |
			Amount	Percent Cover- age[a]	Interest Rate	Repay- ment Terms[b]	Amount
FY 2000 (continued)							
China/Air China	2 x 737, 1 x 747, 2 x 777	$379	$ —	—%	—%	—	$323
China/Hainan Airlines	2 x 737	96	—	—	—	—	82
China/Shanghai Airlines ...	2 x 737	86	—	—	—	—	73
Czech Republic/Czech Airlines	2 x 737	74	—	—	—	—	59
Ireland/Ryanair	5 x 737	164	—	—	—	—	134
Korea/Korean Air Lines ...	—	21	10	47.6	7.05	20-Q	8
Korea/Korean Air Lines ...	3 x 777	384	—	—	—	—	317
Morocco/Royal Air Maroc	3 x 737	108	—	—	—	—	92
Panama/Cia Panamena De Aviacion	4 x 737	149	—	—	—	—	105
Russia/Aeroflot Russian International Airlines......	engines & avionics	129	—	—	—	—	130
Saudi Arabia/Ministry of Finance	1 x 747, 3 x 777	661	—	—	—	—	551
Taiwan/China Airlines	1 x 737, 3 x 747	472	—	—	—	—	403
Taiwan/Eva Airways	1 x 747	144	—	—	—	—	119
Taiwan/Far Eastern Air Transport.....................	2 x 757	101	—	—	—	—	84
Thailand/Thai Airways International	1 x 747, 4 x 777	613	—	—	—	—	503
Tunisia/Societe Tunisienne De L'air	1 x 737	36	—	—	—	—	31
Turkey/Turk Hava Yollari Tao	7 x 737	299	—	—	—	—	221

Source: Aerospace Industries Assocation, based on data from the Export-Import Bank of the United States.
NOTE: For definitions of Loans and Guarantees, see Export-Import Bank tables on previous pages.
   a Amount of loan as percent of export value.
   b Number of payments and frequency (S=semi-annual).

EMPLOYMENT

The aerospace industry, which employed 4.5% of all manufacturing and 7.4% of durable goods manufacturing workers, reduced its workforce in 2001 by 9,000. On an annual average basis, the aerospace industry employed 790,000 workers with an annual payroll of $31 billion. The aircraft sector employed 461,000, while the missiles and space sector employed 84,000 and avionics and other input industries provided another 246,000 jobs.

Of the jobs lost in 2001, the aircraft and parts manufacturing sector accounted for 3,500 of them. Engine sector employment dropped 1,700 while the complete aircraft manufacturing sector cut 1,300 and the aircraft parts manufacturing eased 500. Also declining, missiles and space sector's employment dropped by 2,600 and the avionics and other input industry employment fell by 2,000.

Earnings of production workers increased on average to $21.11 per hour—up from $20.52 in 2000. Production workers doing aircraft final assembly earned the highest hourly wage—$23.83. Final assembly workers in the missiles and space sector were next with $22.60 per hour; followed by aircraft engine workers, $20.64, and aircraft parts, $18.44.

Aircraft engine workers on average work a longer workweek—44.0 hours. This compares with 43.0 in the parts sector, 42.1 for missiles and space primes, and 41.7 for aircraft final assembly. Final assembly workers in the missiles and space sector logged the highest average overtime—10.2 hours per week, up from 8.1 in 2000 and 3.1 in 1999.

The Bureau of Labor Statistics (BLS) reported that the hourly labor cost, including benefits, is $41.75 for all workers in the aircraft manu-facturing industry. This is composed of $27.80 in wages and salaries and $13.95 in benefits. On average, white-collar workers cost $44.98 per hour and blue-collar workers cost $36.56 per hour.

The aerospace industry suffered one work stoppage in 2001. A total of 5,000 employees and 45,000 work-days went idle, according to statistics published by the BLS. The year before, three work-stoppages idled 22,400 workers and 566,400 work-days were lost.

National employment of R&D-performing scientists and engineers across all industries has steadily grown since 1995, while aerospace employment of R&D-performing scientists has steadily declined. In 1996, 95,500 R&D scientists and engineers found employment in the aerospace industry. By 2001, that number had dropped to 25,100. The aerospace share of national employment has fallen to the lowest level on record—just 2.4%. Despite fewer R&D jobs in aerospace, most experts agree that engineers and technical workers remain in demand. Further, most aerospace companies say they still need mechanical, structural, and aerodynamics engineers, as well as software engineers and information technology specialists.

ANNUAL AVERAGE EMPLOYMENT IN ALL MANUFACTURING, DURABLE GOODS, AND AEROSPACE INDUSTRIES
Calendar Years 1979–2001
(Thousands)

Year	All Manu-facturing Industries	Durable Goods Industries	Aerospace Industry[a]		
				As Percent of	
			Total	All Manufac-turing	Durable Goods
1979	21,040	12,730	1,007	4.8%	7.9%
1980	20,285	12,159	1,080	5.3	8.9
1981	20,170	12,082	1,087	5.4	9.0
1982	18,780	11,014	1,038	5.5	9.4
1983	18,432	10,707	1,019	5.5	9.5
1984	19,372	11,476	1,058	5.5	9.2
1985	19,248	11,458	1,151	6.0	10.1
1986	18,947	11,195	1,241	6.6	11.1
1987	18,999	11,154	1,282	6.8	11.5
1988	19,314	11,363	1,294	6.7	11.4
1989	19,391	11,394	1,314	6.8	11.5
1990	19,076	11,109	1,302	6.8	11.7
1991	18,406	10,569	1,214	6.6	11.5
1992	18,104	10,277	1,100	6.1	10.7
1993	18,075	10,221	966	5.3	9.5
1994	18,321	10,448	855	4.7	8.2
1995	18,524	10,683	796	4.3	7.5
1996	18,495	10,789	796	4.3	7.4
1997	18,675	11,010	859	4.6	7.8
1998	18,805	11,205	896	4.8	8.0
1999	18,552	11,111	847	4.6	7.6
2000[r]	18,473	11,141	799	4.3	7.2
2001	17,695	10,636	790	4.5	7.4

Source: Bureau of Labor Statistics and Aerospace Industries Association estimates.
  a See Glossary for detailed explanation of "Aerospace Employment."

ANNUAL PAYROLL
OF ALL MANUFACTURING AND AEROSPACE INDUSTRIES
Calendar Years 1979–2001
(Millions of Dollars)

Year	All Manufacturing Industries[a]	Aerospace Industry[b]			Aerospace As Percent of All Manufacturing
		TOTAL	Production Workers	Other Workers	
1979	$335,200	$15,150	$ 6,465	$ 8,685	4.5%
1980	356,200	18,026	7,658	10,368	5.1
1981	387,600	19,906	8,152	11,754	5.1
1982	385,700	20,750	8,043	12,707	5.4
1983	400,700	21,644	8,071	13,573	5.4
1984	445,400	23,773	8,746	15,027	5.3
1985	468,500	26,749	9,837	16,911	5.7
1986	480,700	29,547	11,038	18,509	6.1
1987	496,900	31,101	11,700	19,401	6.3
1988	529,900	32,566	11,744	20,822	6.1
1989	547,900	34,154	12,440	21,714	6.2
1990	561,400	35,590	13,020	22,570	6.3
1991	562,500	34,520	12,536	21,984	6.1
1992	583,500	33,123	11,812	21,311	5.7
1993	592,400	30,391	10,673	19,718	5.1
1994	620,300	28,395	9,901	18,494	4.6
1995	647,500	26,603	9,272	17,331	4.1
1996	673,700	27,987	10,105	17,882	4.2
1997	718,400	31,575	12,092	19,483	4.4
1998	756,600	32,981	12,769	20,213	4.4
1999	782,000	31,143	11,661	19,481	4.0
2000[r]	830,100	31,160	11,153	20,007	3.8
2001	842,200	31,258	11,155	20,103	3.7

Source: Bureau of Economic Analysis, "Survey of Current Business" (Monthly) and Aerospace Industries Association estimates based on data from the Bureau of Labor Statistics.
a See Glossary for explanation of "Payroll, All Manufacturing."
b Based on combined annual average employment and average weekly earnings for SICs 372 and 376.

EMPLOYMENT IN THE AEROSPACE INDUSTRY[a]
Calendar Years 1987–2001
(Thousands)

Year	TOTAL	Aircraft, Engines, & Parts (SIC 372)	Missiles & Space Vehicles (SIC 376)	Other[b]
TOTAL EMPLOYMENT				
1987	1,282	678	206	399
1988	1,294	684	208	402
1989	1,314	711	194	408
1990	1,302	712	185	405
1991	1,214	669	168	378
1992	1,100	612	146	342
1993	966	542	124	300
1994	855	482	108	266
1995	796	451	98	248
1996	796	458	90	248
1997	859	501	91	267
1998	896	525	92	279
1999	847	496	87	263
2000[r]	799	464	86	248
2001	790	461	84	246
PRODUCTION WORKERS				
1987	434	339	67	29
1988	422	331	63	28
1989	432	344	60	29
1990	430	345	57	29
1991	399	324	48	27
1992	355	291	40	24
1993	308	253	35	20
1994	271	222	31	18
1995	252	208	28	17
1996	260	218	25	17
1997	295	251	25	20
1998	311	266	25	21
1999	285	244	23	19
2000[r]	259	220	22	17
2001	255	219	20	17

Source: Bureau of Labor Statistics and Aerospace Industries Association estimates.
   a Annual average. See Glossary for detailed explanation of "Aerospace Employment."
   b Communications, navigation, flight control, and displays (aerospace-related portions of SICs 366, 381, & 382).

EMPLOYMENT IN THE AIRCRAFT, ENGINES, AND PARTS INDUSTRY[a]
Calendar Years 1987–2001
(Thousands)

Year	TOTAL (SIC 372)	Aircraft (SIC 3721)	Engines and Parts (SIC 3724)	Other Parts & Equipment (SIC 3728)
TOTAL EMPLOYMENT				
1987	678.0	356.4	158.2	163.4
1988	683.5	368.5	155.8	159.3
1989	711.0	382.2	153.5	175.2
1990	712.3	381.0	151.7	179.5
1991	669.2	355.6	143.2	170.3
1992	611.7	332.1	126.6	153.0
1993	542.0	301.4	109.2	131.4
1994	481.5	271.3	95.1	115.1
1995	450.5	243.6	93.0	113.9
1996	458.1	243.1	94.7	120.4
1997	500.6	262.4	99.8	138.4
1998	525.1	271.6	103.3	150.2
1999	496.3	253.9	101.2	141.2
2000[r]	464.1	233.9	100.6	129.6
2001	460.6	232.6	98.9	129.1
PRODUCTION WORKERS				
1987	338.5	159.1	80.5	96.3
1988	331.3	162.1	77.1	92.1
1989	343.7	167.4	76.8	99.5
1990	344.6	164.1	77.2	103.2
1991	323.6	151.6	73.1	98.8
1992	291.4	137.8	64.3	89.2
1993	252.5	122.7	53.6	76.2
1994	222.0	108.1	46.9	67.0
1995	207.5	93.6	46.2	67.7
1996	217.7	95.6	48.8	73.3
1997	251.1	110.1	53.6	87.4
1998	265.5	114.9	54.0	96.5
1999	243.8	105.2	49.3	89.3
2000[r]	220.1	91.3	48.1	80.7
2001	218.5	88.8	48.4	81.3

Source: Bureau of Labor Statistics.
  a   Annual average. See Glossary for detailed explanation of "Aerospace Employment."

AVERAGE WEEKLY EARNINGS IN THE AEROSPACE INDUSTRY
Production Workers Only
Calendar Years 1979–2001

| Year | TOTAL[a] | Aircraft and Parts (SIC 372) | | | | Guided Missiles, Space Vehicles & Parts (SIC 376) | Complete Guided Missiles, & Space Vehicles (SIC 3761) |
		TOTAL[a]	Aircraft (SIC 3721)	Engines & Parts (SIC 3724)	Other Parts & Equipment (SIC 3728)		
AVERAGE WEEKLY EARNINGS[b]							
1979	$351	$351	$360	$361	$322	$347	$348
1980	389	390	404	394	358	378	383
1981	424	426	444	422	396	410	420
1982	460	462	485	454	426	447	461
1983	486	487	513	476	453	480	494
1984	513	516	532	523	486	496	508
1985	531	534	547	542	506	515	527
1986	545	550	568	561	520	517	533
1987	556	558	578	567	523	541	556
1988	573	575	596	582	529	567	585
1989	593	594	616	616	542	589	611
1990	624	626	656	637	570	612	634
1991	648	651	694	654	583	632	649
1992	685	689	736	689	615	652	666
1993	714	717	756	715	657	696	727
1994	754	756	800	753	688	738	779
1995	758	757	809	770	677	765	812
1996	801	802	859	813	721	790	837
1997	844	844	918	838	756	842	896
1998	847	847	932	840	751	840	892
1999	842	841	923	859	731	852	896
2000	888[r]	892[r]	993	898	774	847	875
2001	900	900	994	908	793	901	951

Source: Bureau of Labor Statistics and Aerospace Industries Association estimates.
   a TOTAL columns are employment-based weighted averages.
   b Includes overtime premiums.

AVERAGE HOURLY EARNINGS IN THE AEROSPACE INDUSTRY
Production Workers Only
Calendar Years 1979–2001

| Year | TOTAL[a] | Aircraft and Parts (SIC 372) | | | | Guided Missiles, Space Vehicles & Parts (SIC 376) | Complete Guided Missiles, & Space Vehicles (SIC 3761) |
		TOTAL[a]	Aircraft (SIC 3721)	Engines & Parts (SIC 3724)	Other Parts & Equipment (SIC 3728)		
AVERAGE HOURLY EARNINGS[b]							
1979	$ 8.26	$ 8.26	$ 8.50	$ 8.53	$ 7.48	$ 8.25	$ 8.38
1980	9.27	9.28	9.66	9.42	8.40	9.22	9.33
1981	10.29	10.31	10.74	10.41	9.35	10.06	10.34
1982	11.20	11.23	11.85	11.16	10.17	10.95	11.21
1983	11.79	11.82	12.58	11.61	10.73	11.59	11.84
1984	12.24	12.32	12.91	12.40	11.37	11.82	12.01
1985	12.54	12.62	13.18	12.85	11.66	12.14	12.36
1986	12.75	12.86	13.48	13.08	11.90	12.20	12.48
1987	13.10	13.17	13.74	13.33	12.23	12.73	13.09
1988	13.48	13.55	14.18	13.80	12.28	13.13	13.53
1989	14.10	14.17	14.89	14.42	12.81	13.70	14.20
1990	14.73	14.79	15.66	14.84	13.37	14.39	14.82
1991	15.51	15.60	16.72	15.38	14.05	14.90	15.21
1992	16.46	16.53	17.70	16.28	14.89	15.99	16.45
1993	17.18	17.23	18.43	16.70	15.72	16.80	17.43
1994	17.89	17.95	19.50	17.31	16.01	17.48	18.29
1995	17.99	18.02	19.97	17.34	15.93	17.74	18.58
1996	18.56	18.57	20.49	18.22	16.42	18.51	19.34
1997	18.94	18.88	20.76	18.58	16.76	19.53	20.75
1998	19.24	19.17	21.08	18.93	17.02	19.96	21.38
1999	19.67[r]	19.60	21.83	19.47	17.08	20.39	21.96
2000	20.52[r]	20.50[r]	23.14	20.17	17.76	20.76	21.98
2001	21.11	21.08	23.83	20.64	18.44	21.40	22.60

Source: Bureau of Labor Statistics and Aerospace Industries Association estimates.
   a   TOTAL columns are employment-based weighted averages.
   b   Includes overtime premiums.

AVERAGE HOURS IN THE AEROSPACE INDUSTRY
Production Workers Only
Calendar Years 1987–2001

| Year | TOTAL[a] | Aircraft and Parts (SIC 372) | | | | Guided Missiles, Space Vehicles & Parts (SIC 376) | Complete Guided Missiles, & Space Vehicles (SIC 3761) |
		TOTAL[a]	Aircraft (SIC 3721)	Engines & Parts (SIC 3724)	Other Parts & Equipment (SIC 3728)		
AVERAGE WEEKLY HOURS							
1987	42.4	42.4	42.1	42.5	42.8	42.5	42.5
1988	42.5	42.4	42.0	42.2	43.1	43.2	43.2
1989	42.1	41.9	41.4	42.7	42.3	43.0	43.0
1990	42.3	42.3	41.9	42.9	42.6	42.5	42.8
1991	41.8	41.7	41.5	42.5	41.5	42.4	42.7
1992	41.6	41.7	41.6	42.3	41.3	40.8	40.5
1993	41.6	41.6	41.0	42.8	41.8	41.4	41.7
1994	42.1	42.1	41.0	43.5	43.0	42.2	42.6
1995	42.1	42.0	40.5	44.4	42.5	43.1	43.7
1996	43.1	43.2	41.9	44.6	43.9	42.7	43.3
1997	44.6	44.7	44.2	45.1	45.1	43.1	43.2
1998	44.0	44.2	44.2	44.4	44.1	42.1	41.7
1999	42.8	42.9	42.3	44.1	42.8	41.8	40.8
2000	43.3	43.5	42.9	44.5	43.6	40.8	39.8
2001	42.7	42.7	41.7	44.0	43.0	42.1	42.1
AVERAGE WEEKLY OVERTIME HOURS							
1987	4.8	4.9	4.4	5.0	5.4	4.2	4.3
1988	4.6	4.6	4.3	4.6	5.1	4.5	4.6
1989	5.0	5.1	5.0	5.4	5.0	4.4	4.5
1990	4.5	4.6	4.3	5.3	4.5	3.8	4.1
1991	4.0	4.0	4.1	4.5	3.5	3.9	4.5
1992	3.6	3.7	3.6	4.4	3.3	2.8	3.1
1993	3.8	3.9	3.7	4.6	3.7	2.9	3.2
1994	4.5	4.6	4.1	5.3	4.8	3.7	3.8
1995	4.8	4.9	4.2	5.9	5.2	4.2	4.6
1996	5.7	5.9	5.3	6.5	6.3	3.9	4.2
1997	6.9	7.2	7.2	6.8	7.3	4.3	4.3
1998	5.9	6.1	5.9	6.0	6.3	3.8	3.6
1999	4.4	4.5	4.3	5.4	4.3	3.5	3.1
2000	4.9	4.8	4.6	6.0	4.4	5.7	8.1
2001	4.8	4.6	4.1	6.0	4.4	6.5	10.2

Source: Bureau of Labor Statistics and Aerospace Industries Association estimates.
a TOTAL columns are employment-based weighted averages.

EMPLOYER COSTS FOR EMPLOYEE COMPENSATION IN THE AIRCRAFT MANUFACTURING INDUSTRY

March[a] 1997–2002

	1997	1998	1999	2000	2001	2002
ALL OCCUPATIONS						
TOTAL	$33.98	$34.27	$35.33	$37.87	$40.09	$41.75
Wages and salaries	22.63	23.32	24.28	25.47	26.79	27.80
Benefits—TOTAL	11.35	10.95	11.05	12.39	13.30	13.95
Paid leave	3.07	3.23	3.34	3.51	3.69	3.82
Supplemental pay	1.05	1.08	1.19	1.67	1.79	1.78
Insurance	2.75	2.48	2.60	2.67	3.13	3.51
Retirement & savings	1.58	1.42	0.97	1.20	1.36	1.41
Legally required	2.81	2.69	2.77	2.92	3.12	3.20
Other	0.09	0.04	0.18	0.42	0.21	0.22
WHITE-COLLAR OCCUPATIONS						
TOTAL	$36.32	$36.97	$38.12	$40.76	$43.63	$44.98
Wages and salaries	24.83	25.68	26.88	28.31	29.68	30.64
Benefits—TOTAL	11.50	11.30	11.24	12.44	13.95	14.34
Paid leave	3.43	3.52	3.71	4.04	4.22	4.36
Supplemental pay	0.72	0.77	0.71	0.81	1.56	1.28
Insurance	2.66	2.48	2.60	2.69	3.19	3.60
Retirement & savings	1.71	1.61	1.12	1.32	1.50	1.56
Legally required	2.90	2.88	2.87	2.99	3.23	3.30
Other	0.08	0.04	0.24	0.59	0.26	0.23
BLUE-COLLAR OCCUPATIONS						
TOTAL	$30.14	$29.56	$30.56	$33.33	$34.18	$36.56
Wages and salaries	19.04	19.24	19.84	20.94	22.00	23.26
Benefits—TOTAL	11.10	10.32	10.72	12.39	12.18	13.30
Paid leave	2.48	2.72	2.70	2.66	2.81	2.95
Supplemental pay	1.64	1.63	2.02	3.13	2.18	2.59
Insurance	2.91	2.49	2.60	2.61	3.02	3.35
Retirement & savings	1.37	1.08	0.73	1.02	1.10	1.15
Legally required	2.66	2.35	2.58	2.83	2.94	3.06
Other	0.05	0.05	0.09	0.13	0.13	0.21

Source: Bureau of Labor Statistics, "Employer Costs for Employee Compensation" (Annually).
a Based on the pay period including March 12th.

WORK STOPPAGES IN THE AEROSPACE INDUSTRY
Calendar Years 1979–2001

Year	Number of Strikes[a]	Number of Workers Involved	Work-Days Idle in Year
1979	12	6,600	103,400
1980	17	4,400	92,900
1981	12	6,100	188,900
1982[b]	4	11,900	45,200
1983	2	8,700	404,100
1984	4	14,600	188,200
1985	4	19,700	289,800
1986	—	—	—
1987	—	—	—
1988	3	10,600	415,800
1989	2	58,500	1,848,000
1990	1	2,300	56,700
1991	1	1,500	—
1992	1	3,800	11,400
1993	2	27,800	34,600
1994	—	—	—
1995	1	33,000	1,551,000
1996	2	7,800	90,100
1997	—	—	—
1998	—	—	—
1999	—	—	—
2000	3	22,400	566,400
2001	1	5,000	45,000

Source: Bureau of Labor Statistics, "Compensation and Working Conditions" (Quarterly).
 a Strikes beginning during calendar year.
 b Effective 1982, data not available for work stoppages involving fewer than 1,000 employees.

OCCUPATIONAL INJURY AND ILLNESS INCIDENCE RATES[a]
ALL MANUFACTURING AND AEROSPACE INDUSTRIES
Calendar Years 1996–2000

Manufacturing Sector	1996	1997	1998	1999	2000
All Manufacturing:					
Total Cases	10.6	10.3	9.7	9.2	9.0
Lost Workday Cases	4.9	4.8	4.7	4.6	4.5
Nonfatal Cases without Lost Workdays	5.7	5.4	5.0	4.6	4.5
Aircraft and Parts (SIC 372):					
Total Cases	7.9	8.7	8.7	8.2	7.1
Lost Workday Cases	3.4	3.8	4.2	4.1	3.4
Nonfatal Cases without Lost Workdays	4.5	4.9	4.5	4.1	3.7
Aircraft (SIC 3721):					
Total Cases	7.3	8.5	8.9	8.9	7.5
Lost Workday Cases	3.0	3.7	4.3	4.4	3.7
Nonfatal Cases without Lost Workdays	4.3	4.8	4.6	4.5	3.7
Aircraft Engines and Parts (SIC 3724):					
Total Cases	7.9	7.7	6.2	5.8	5.5
Lost Workday Cases	3.6	3.3	2.8	2.9	2.2
Nonfatal Cases without Lost Workdays	4.3	4.5	3.4	2.9	3.3
Aircraft Parts (SIC 3728):					
Total Cases	9.1	9.9	10.0	8.7	7.6
Lost Workday Cases	3.9	4.5	5.0	4.4	3.7
Nonfatal Cases without Lost Workdays	5.2	5.4	5.0	4.3	3.9
Guided Missiles, Space Vehicles & Parts (SIC 376):					
Total Cases	3.4	3.2	3.3	2.8	2.2
Lost Workday Cases	1.3	1.5	1.5	1.3	1.1
Nonfatal Cases without Lost Workdays	2.0	1.7	1.8	1.5	1.1
Guided Missiles & Space Vehicles (SIC 3761):					
Total Cases	3.0	2.9	3.0	2.7	2.0
Lost Workday Cases	1.2	1.3	1.3	1.3	1.0
Nonfatal Cases without Lost Workdays	1.8	1.6	1.8	1.4	1.0
Space Propulsion Units & Parts (SIC 3764):					
Total Cases	NA	NA	3.2	NA	3.0
Lost Workday Cases	NA	NA	1.6	NA	1.5
Nonfatal Cases without Lost Workdays	NA	NA	1.7	NA	1.6
Other Space Vehicle Equipment (SIC 3769):					
Total Cases	NA	NA	4.3	3.5	2.2
Lost Workday Cases	NA	NA	2.6	1.6	1.1
Nonfatal Cases without Lost Workdays	NA	NA	1.8	2.0	1.1

Source: Bureau of Labor Statistics, "Survey of Occupational Injuries and Illnesses" (Annually).
   a Defined as the number of injuries and illnesses per 100 full-time workers. Separate incidence rates also available for occupational injuries only.

EMPLOYMENT IN NATIONAL AERONAUTICS
AND SPACE ADMINISTRATION PROGRAMS
End of Fiscal Years 1964–2003

Year	TOTAL	NASA Employees	Contractor Employees [a]
1964	379,084	31,984	347,100
1965	409,900	33,200	376,700
1966	393,924	33,924	360,000
1967	306,926	33,726	273,200
1968	267,871	32,471	235,400
1969	218,345	31,745	186,600
1970	160,850	31,350	129,500
1971	143,578	29,478	114,100
1972	138,800	27,500	111,300
1973	134,850	26,850	108,000
1974	125,220	25,020	100,200
1975	127,733	24,333	103,400
1976	130,739	24,039	108,000
1977	124,136	23,636	100,500
1978	124,637	23,237	101,400
1979	131,931	22,831	109,100
1980	135,613	22,613	113,000
1981	133,473	21,873	111,600
1982	128,730	22,430	106,300
1983	129,246	22,246	107,000
1984	162,080	22,080	140,000
1985	131,991	21,991	110,000
1986	154,660	21,660	133,000
1987	165,001	22,001	143,000
1988	172,326	22,326	150,000
1989	213,054	23,054	190,000
1990	221,829	23,829	198,000
1991	223,149	24,149	199,000
1992	230,513	24,513	206,000
1993	228,674	24,174	204,500
1994	217,910	23,873	194,037
1995	209,355	22,355	187,000
1996	198,113	21,113	177,000
1997	189,070	20,070	169,000
1998	183,109	19,109	164,000
1999	181,469	18,469	163,000
2000	173,375	18,375	155,000
2001	171,700	18,700	153,000
2002 [E]	174,000	19,000	155,000
2003 [E]	177,100	19,100	158,000

Source: Office of Management and Budget, "Budget of the United States Government" (Annually) and NASA Headquarters.

a Includes estimates of manpower for hardware and related contracts, as well as actual work-years for support service contracts. Increase in FY 1984 caused by change in estimating methodology to reflect more accurately the mix of support and development contractors.

FEDERAL CIVILIAN EMPLOYMENT[a] IN THE DEPARTMENT OF DEFENSE
Fiscal Years 1967–2003

Year	TOTAL	Civil Functions[b]	Military Functions[c]
1967	1,225,637	31,980	1,193,657
1968	1,288,130	32,062	1,256,068
1969	1,257,091	31,214	1,225,877
1970	1,159,935	30,293	1,129,642
1971	1,092,804	30,063	1,062,741
1972	1,040,147	30,585	1,009,562
1973	987,281	29,971	957,310
1974	1,002,850	29,072	973,778
1975	983,790	29,069	954,721
1976	951,034	28,648	922,386
1977	940,549	28,912	911,637
1978	933,071	28,962	904,109
1979	914,582	28,592	885,990
1980	907,700	27,700	880,000
1981	981,400	34,400	947,000
1982	1,009,192	31,111	978,081
1983	1,015,622	30,816	984,806
1984	1,040,213	28,681	1,011,532
1985	1,065,624	28,754	1,036,870
1986	1,069,863	28,511	1,041,352
1987	1,059,669	28,352	1,031,317
1988	1,053,000	28,419	1,024,581
1989	1,051,166	28,081	1,023,085
1990	1,048,814	27,651	1,021,163
1991	1,001,183	27,385	973,798
1992	1,000,453	27,584	972,869
1993	958,855	27,055	931,800
1994	896,293	28,001	868,292
1995	849,529	27,790	821,739
1996	806,122	27,180	778,942
1997	771,914	26,164	745,750
1998	732,097	24,855	707,242
1999	705,826	24,830	680,996
2000	685,163	24,878	660,285
2001	674,811	24,920	649,891
2002 [E]	659,533	24,800	634,733
2003 [E]	650,630	23,200	627,430

Source: Office of Management and Budget, "The Budget of the United States Government" (Annually).
  a   Full-time equivalent civilian employment.
  b   Data are estimated for portions of Civil Functions.
  c   The Department of Defense is exempt from full-time equivalent controls. Data shown are estimated civilian employment for military functions and military assistance.

FINANCE

The aerospace industry generated $6.6 billion in net income after taxes on $169 billion of corporate sales last year. Net profit as a percentage of sales fell to 3.9%—the lowest level since 1995—down from 4.7% in 2000 and 6.5% in 1999. The corresponding profit margin for all manufacturing corporations was 0.8%—down sharply from 2000's 6.1%. Similarly, the aerospace industry's net profit as a percentage of assets and shareholders' equity declined to 3.6% and 11.6%, respectively. For comparison, prior year aerospace returns were 4.3% and 14.2%; and the averages for all manufacturing corporations in 2001 were 0.8% and 1.9%, respectively. Industry-wide working capital declined marginally after falling $2.1 billion in 2000.

Perhaps not coincidentally, the aerospace industry's capital equipment expenditures declined 32% to $2.3 billion in 2000, the latest data available. Missile/space sector investment fell sharply to $432 million after rising substantially over the last four years. Meanwhile, the aircraft sector's capital equipment investment declined $447 million to $1.9 billion.

DoD awarded $33 billion in aircraft procurement contracts in FY 2001. The South Atlantic region received $1.9 billion more aircraft procurement contracts than it had in the previous year. West South Central, however, did much worse—booking $2.7 billion fewer orders than last year. Awards for missile and space systems fell slightly to $9.3 billion; and electronics and communications equipment procurement jumped $2 billion to $14 billion. Companies in the Pacific region received the most missile and space systems contracts, $3.9 billion, while companies in the South Atlantic region garnered the most electronics and communications equipment contracts, $4.4 billion.

Lockheed Martin retained its position as the DoD's largest contractor in FY 2001—winning $15 billion in prime contract awards or about 10% of all the available DoD prime contract dollars that year.

The company's contract wins were 3% lower than in 2000; and it was the fifth year in a row that Lockheed Martin topped the DoD list. Boeing made the second spot for the fifth straight year—winning $13 billion worth of prime contracts. Rounding out the top ten DoD contractors were: Newport News Shipbuilding, $5.9 billion; Raytheon, $5.6 billion; Northrop Grumman, $5.2 billion; General Dynamics, $4.9 billion; United Technologies, $3.4 billion; TRW, $1.9 billion; Science Applications International, $1.7 billion; and General Electric, $1.7 billion. Together the top ten contractors accounted for 40% of the 2001 total prime contract dollars awarded or more than $58 billion.

United Space Alliance topped NASA's contractor list in FY 2001 with awards totaling $1.7 billion. The Boeing Company ranked second with $952 million. At $608 million, Lockheed Martin captured third. Other subsidiaries and affiliates of Lockheed Martin and The Boeing Company won additional contracts.

INCOME STATEMENT AND OPERATING RATIOS
FOR AEROSPACE COMPANIES[a]
Calendar Years 1998–2001
(Millions of Dollars)

INCOME STATEMENT	1998	1999	2000	2001
Net Sales, Receipts, Operating Revenues	$154,606	$157,087	$154,877	$168,756
Less: Depreciation, Depletion, & Amortization of Property, Plant, and Equipment	4,201	4,212	4,330	3,895
Less: All Other Operating Costs & Expenses, including Selling Costs & General & Administrative Expenses	139,118	140,228	137,714	152,182
Income (or Loss) from Operations	$ 11,287	$ 12,648	$ 12,833	$ 12,679
Net Non-Operating Income (Expense)	(431)	1,775	(1,886)	(4,059)
Income (or Loss) before Income Taxes (= Total Income)	$ 10,855	$ 14,423	$ 10,947	$ 8,618
Less: Provision for Current & Deferred Domestic Income Taxes	3,155	4,208	3,686	2,054
Income (or Loss) after Income Taxes (= Net Profit)	$ 7,701	$ 10,214	$ 7,260	$ 6,565
Cash Dividends Charged to Retained Earnings	2,397	2,501	2,684	2,561
Net Income Retained in Business	$ 5,304	$ 7,713	$ 4,576	$ 4,003
Retained Earnings at Beginning of Year[b]	31,130	34,832	44,949	52,966
Adjustments to Retained Earnings[c]	(42)	(607)	(633)	(1,597)
Retained Earnings at End of Year[d]	$ 36,392	$ 41,938	$ 48,892	$ 55,373

OPERATING RATIOS

	1998	1999	2000	2001
Income before Taxes as Percent of Net Sales	7.0%	9.2%	7.1%	5.1%
Provision for Current & Deferred Domestic Income Taxes as Percent of Income before Taxes (Total Income)	29.1	29.2	33.7	23.8
Income after Taxes (Net Profit) as Percent of Net Sales	5.0	6.5	4.7	3.9
Income after Taxes (Net Profit) as Percent of Stockholders' Equity[e]	18.0	21.8	14.2	11.6
Income after Taxes (Net Profit) as Percent of Total Assets[e]	4.8	6.2	4.3	3.6

Source: Bureau of the Census, "Quarterly Financial Report for Manufacturing, Mining, and Trade Corporations" (Quarterly).
   a   Based on sample of corporate entities classified in NAICS code 3364, having as their principal activity the manufacture of aerospace products and parts. Prior to 2001, data categorized using SIC system and reported combining codes 372 and 376.
   b   Beginning-of-year retained earnings for any particular year do not equal end-of-year retained earnings for the previous year because of rotation of small companies in survey sample.
   c   Other direct credits (or charges) to retained earnings (net), including stock and other non-cash dividends, etc.
   d   Retained Earnings at End of Year CALCULATED AS Retained Earnings at Beginning of Year PLUS Income (Loss) after Income Taxes MINUS Cash Dividends Charged to Retained Earnings PLUS Adjustments to Retained Earnings.
   e   Average of four quarters.

BALANCE SHEET FOR AEROSPACE COMPANIES[a]
As of December 31, 1998–2001
(Millions of Dollars)

	1998	1999	2000	2001
Assets:				
Current Assets:				
Cash	$ 1,918	$ 2,898	$ 2,969	$ 3,004
Securities, Commercial Paper, & Other				
Short-term Financial Investments	2,364	3,165	1,353	638
Total Cash and U.S. Government				
and Other Securities	$ 4,283	$ 6,063	$ 4,322	$ 3,642
Receivables (Total)	16,765	18,082	20,283	19,438
Inventories (Gross)	46,578	41,216	41,683	42,343
Other Current Assets	7,730	8,320	8,987	10,450
Current Assets—TOTAL	$ 75,356	$ 73,681	$ 75,275	$ 75,872
Net Plant, Property, & Equipment	26,721	28,613	28,274	28,722
Other Non-Current Assets	57,779	67,701	70,908	79,410
Assets—TOTAL	$159,856	$169,995	$174,457	$184,004
Liabilities:				
Current Liabilities:				
Short Term Loans	$ 4,178	$ 4,009	$ 3,901	$ 2,084
Trade Accounts & Notes Payable	11,634	11,001	11,758	10,458
Income Taxes Accrued	2,429	1,987	3,609	1,778
Installments Due on Long Term Debts	2,098	1,793	2,772	2,534
Other Current Liabilities	43,524	43,204	43,668	49,708
Current Liabilities—TOTAL	$ 63,862	$ 61,994	$ 65,709	$ 66,562
Long Term Debt	28,937	33,485	32,418	33,882
Other Non-Current Liabilities	23,987	24,173	23,840	27,054
Liabilities—TOTAL	$116,787	$119,652	$121,967	$127,497
Stockholders' Equity:				
Capital Stock	$ 8,027	$ 5,682	$ 3,276	$ 4,190
Retained Earnings	35,043	44,661	49,214	52,318
Stockholders' Equity—TOTAL	$ 43,069	$ 50,343	$ 52,490	$ 56,508
Liabilities & Stockholders' Equity—TOTAL	$159,856	$169,995	$174,457	$184,004
Net Working Capital	$ 11,494	$ 11,688	$ 9,566	$ 9,311

Source: Bureau of the Census, "Quarterly Financial Report for Manufacturing, Mining, and Trade Corporations" (Quarterly).
a Based on sample of corporate entities classified in NAICS code 3364, having as their principal activity the manufacture of aerospace products and parts. Prior to 2001, data categorized using SIC system and reported combining codes 372 and 376.

NET PROFIT AFTER TAXES
AS A PERCENT OF SALES, ASSETS, AND EQUITY
FOR ALL MANUFACTURING CORPORATIONS
AND THE AEROSPACE INDUSTRY
Calendar Years 1987–2001

PERCENT OF SALES

Year	All Manufacturing Corporations	Non-Durable Goods	Durable Goods	Aerospace[a] Industry
1987	4.9%	5.2%	4.5%	4.1%
1988	6.0	6.7	5.2	4.3
1989	5.0	5.8	4.1	3.3
1990	4.0	4.9	3.0	3.4
1991	2.5	4.2	0.6	1.8[b]
1992	1.0	3.2	(1.4)	(1.4)[b]
1993	2.8	3.7	1.9	3.6
1994	5.4	5.5	5.2	4.7
1995	5.7	6.1	5.3	3.8
1996	6.0	6.6	5.5	5.6
1997	6.2	6.6	5.8	5.2
1998	6.0	6.1	5.9	5.0
1999	6.2	6.2	6.2	6.5
2000	6.1	6.9	5.4	4.7
2001	0.8	5.7	(3.3)	3.9

Year	Percent of Assets[c]		Percent of Equity[c]	
	All Manufacturing	Aerospace[a] Industry	All Manufacturing	Aerospace[a] Industry
1987	5.6%	4.4%	12.8%	14.6%
1988	6.9	4.4	16.2	14.9
1989	5.6	3.3	13.7	10.7
1990	4.3	3.4	10.7	11.5
1991	2.6	1.9[b]	6.4	6.1[b]
1992	1.0	(1.2)[b]	2.6	(5.2)[b]
1993	2.9	3.5	8.1	13.2
1994	5.8	4.3	15.6	14.8
1995	6.2	3.5	16.2	11.1
1996	6.5	5.1	16.8	17.1
1997	6.6	4.8	16.6	17.3
1998	6.1	4.8	15.7	18.0
1999	6.1	6.2	16.5	21.8
2000	5.9	4.3	15.2	14.2
2001	0.8	3.6	1.9	11.6

Source: Bureau of the Census, "Quarterly Financial Report for Manufacturing, Mining, and Trade Corporations" (Quarterly).
 a Based on a sample of corporate entities classified in NAICS code 3364, having as their principal activity the manufacture of aerospace products and parts. Prior to 2001, data categorized using SIC system and reported combining codes 372 and 376.
 b Reflects unusually large non-operating expenses totalling $3.4 and $8.7 billion in 1991 and 1992, respectively, due to restructuring changes and the implementation of a change in accounting for future retirement benefit costs.
 c Average of four quarters
 ( ) Net loss after taxes.

CAPITAL EXPENDITURES
Calendar Years 1967–2000
(Millions of Dollars)

Year	All Manufacturing Industries	Aerospace Industry[a]	Aircraft, Engines, & Parts	Missiles, Space Vehicles, & Parts
1967	$ 21,503	$ 520	$ 408	$ 111
1968	20,613	399	282	117
1969	22,291	429	340	89
1970	22,164	244	181	62
1971	20,941	115	59	56
1972	24,073	261	169	92
1973	26,979	362	258	104
1974	35,696	407	283	124
1975	37,262	478	369	109
1976	40,545	557	431	126
1977	47,459	673	508	164
1978	55,209	948	775	174
1979	61,533	1,551	1,301	250
1980	70,113	1,923	1,618	306
1981	78,632	2,006	1,637	369
1982	74,562	2,142	1,680	462
1983	61,931	2,159	1,530	629
1984	75,186	3,050	2,091	960
1985	83,058	3,784	2,429	1,356
1986	76,355	4,145	2,818	1,327
1987	78,650	3,612	2,536	1,075
1988	81,593	3,388	2,362	1,026
1989	98,738	3,921	2,800	1,121
1990	105,018	3,490	2,621	869
1991	103,003	3,407	2,823	584
1992	103,188	3,860	3,384	476
1993	103,133	2,725	2,307	418
1994	112,784	2,363	1,969	395
1995	128,473	2,114	1,734	380
1996	139,323	2,513	2,023	490
1997[b]	151,511	3,132	2,380	752
1998	152,708	3,477	2,613	864
1999[r]	150,325	3,422	2,338	1,084
2000	154,917	2,323	1,891	432

Source: Bureau of the Census, "Annual Survey of Manufactures" (Annually).
  a Combined total for establishments in Aircraft, Missiles, Space Vehicles, and Parts Manufacturing.
  b Prior to 1997, figures included only new capital expenditures.

KEY OPERATING COSTS FOR SELECTED AEROSPACE MANUFACTURING CENTERS
As of 2002

State	Location	Total Annual Operating Cost[a] (in Millions)	Hourly Labor Cost		
			Manufacturing	Technical	Clerical
AZ	Phoenix/Tempe	$33.86	$19.79	$32.65	$13.83
AZ	Tucson................................	32.05	18.87	31.12	13.18
CA	Chula Vista	36.72	21.12	34.84	14.10
CA	El Segundo/Long Beach/ Redondo Beach/Seal Beach/ Torrance	38.88	22.51	37.14	15.73
CA	Sunnyvale	42.77	23.34	38.50	16.30
CO	Denver................................	35.20	20.78	34.28	14.51
CT	Hartford/Stratford/Windsor Locks	36.65	21.85	36.04	15.26
FL	Melbourne	30.75	18.79	30.99	13.12
GA	Marietta	34.19	20.04	33.05	13.99
GA	Savannah............................	31.13	19.17	31.62	13.39
IA	Cedar Rapids	32.76	19.77	32.62	13.81
IL	Rockford	34.07	20.44	33.71	14.28
IN	Indianapolis	34.69	19.80	33.08	14.01
KS	Wichita	33.71	19.93	32.88	13.92
MA	Lexington/Marlborough...........	35.04	20.76	34.25	14.50
MN	Hopkins/Minneapolis	35.48	21.10	34.81	14.74
MO	St. Louis	35.26	20.66	34.08	14.43
NC	Charlotte	34.06	19.79	32.65	13.83
NH	Nashua	35.15	20.76	34.25	14.50
OH	Cleveland	35.24	20.46	33.75	14.29
OH	Cincinnatti/Evandale	34.15	20.20	33.32	14.11
PA	King of Prussia	36.46	21.35	35.21	14.91
SC	Greenville	30.68	18.85	31.09	13.16
TX	Dallas/Ft. Worth/Grand Prairie/ Irving.................................	32.76	19.85	32.75	13.87
UT	Magna	33.31	19.63	32.38	13.71
WA	Auburn/Everett/Renton	36.13	21.18	34.94	14.80

(Continued on next page)

KEY OPERATING COSTS FOR SELECTED AEROSPACE
MANUFACTURING CENTERS
As of 2002, continued

Power (¢/kwh)	Lease Rates ($/sq ft) Industrial	Office	Construction ($/sq ft)	State	Location
6.39¢	$ 4.65	$24.30	$47.79	AZ	Phoenix/Tempe
8.09	6.00	23.00	46.92	AZ	Tucson
12.18	7.56	30.35	57.18	CA	Chula Vista
14.21	7.35	28.15	58.23	CA	El Segundo/Long Beach/ Redondo Beach/Seal Beach/ Torrance
13.73	16.85	49.50	64.14	CA	Sunnyvale
4.92	2.25	21.00	47.48	CO	Denver
7.15	4.85	26.50	57.86	CT	Hartford/Stratford/Windsor Locks
6.84	3.90	17.75	49.82	FL	Melbourne
6.27	4.60	22.90	45.18	GA	Marietta
6.27	3.50	14.50	41.76	GA	Savannah
4.17	3.80	16.75	48.48	IA	Cedar Rapids
7.15	4.10	15.25	58.30	IL	Rockford
4.96	8.00	19.60	51.03	IN	Indianapolis
5.31	4.00	16.50	45.18	KS	Wichita
9.88	4.95	27.25	57.55	MA	Lexington/Marlborough
5.35	3.25	14.00	61.90	MN	Hopkins/Minneapolis
5.07	3.00	24.00	57.36	MO	St. Louis
5.08	6.50	22.50	41.83	NC	Charlotte
8.89	7.10	17.20	51.46	NH	Nashua
9.35	4.00	21.20	55.00	OH	Cleveland
5.98	3.25	20.10	50.09	OH	Cincinnatti/Evandale
8.91	3.75	27.00	62.15	PA	King of Prussia
5.08	3.40	18.25	42.39	SC	Greenville
7.31	3.75	19.35	44.81	TX	Dallas/Ft. Worth/Grand Prairie/ Irving
4.78	4.20	19.20	47.86	UT	Magna
5.42	5.10	22.50	57.09	WA	Auburn/Everett/Renton

Source: The Boyd Company (Princeton, NJ), BizCosts ® data bank.
   a   Includes all major geographically-variable operating costs for a representative 200,000 sq. ft. aerospace manufacturing facility. Costs reflect a 500-worker plant site having 250 production employees, 150 engineers, and 100 administrative support workers. Annual costs for electric power, real estate, and construction are scaled accordingly.

DEPARTMENT OF DEFENSE MAJOR CONTRACTORS
Fiscal Years 1997–2001
**Listed by rank according to net value of
prime contracts awarded during last fiscal year
(Millions of Dollars)**

Company	1997	1998	1999	2000	2001
TOTAL CONTRACT AWARDS	$116,680	$118,139	$125,037	$133,232	$144,635
Lockheed Martin Corp.	$ 11,638	$ 12,341	$ 12,675	$ 15,126	$ 14,687
The Boeing Co.	9,645[b]	10,866[b]	11,568[b]	12,041	13,341
Newport News Shipbuilding Inc. ...	720	1,546	535	790	5,889
Raytheon Co.	5,693	5,661	6,401	6,331	5,576
Northrop Grumman Corp.[c]	5,079	4,335	5,290	5,817	5,153
General Dynamics Corp.	3,012	3,680	4,564	4,196	4,907
United Technologies Corp.	1,810	1,983	2,368	2,072	3,373
TRW Inc....................................	1,163[d]	1,346	1,431	2,005	1,903
Science Applications Int'l Corp.	1,095	1,224	1,358	1,522	1,748
General Electric Co.	1,677	1,161	1,714	1,609	1,747
Carlyle Group.............................	611[f]	1,329	1,336	1,195	1,232
Health Net Inc.	(a)	(a)	(a)	551	939
Dyncorp	535	537	566	771	909
Honeywell International Inc.	547[g]	655[g]	746[g]	951	902
BAE Systems PLC	247[h]	732[h]	729[h]	997	865
GM GDLS Defense Group	(a)	(a)	(a)	(a)	840
Computer Sciences Corp.	704	647	744	1,165	819
ITT Industries Inc.	790	781	659	554	808
Exxon Mobil Corp.	539[i]	381[i]	208[i]	325	705
Bechtel Corp.	267	201	600	695	621
IT Group Inc.	(a)	436	459	493	587
Textron Inc.	1,445	1,838	1,423	1,164	572
Oshkosh Truck Corp.	205	137	253	373	558
Triwest Healthcare Alliance Co.	213	420	414	336	554
Boeing Sikorsky Comanche Team ...	168	197	296	385	528
L-3 Communications Holding, Inc....	(a)	225	316	378	495
Alliant Techsystems Inc.	378	317	422	470	494
B P Amoco PLC	(a)	(a)	(a)	592	489
Worldcom, Inc.	226[j]	235[j]	(a)	(a)	482
Equilon Enterprises LLC.................	(a)	(a)	203	266	466

Source: Department of Defense, "100 Companies Receiving the Largest Dollar Volume of Prime Contract Awards" (Annually).
  a Not in top 100 companies for indicated year(s).
  b Includes awards previously reported separately as Rockwell International Corp.
  c Includes awards previously reported separately as Litton Industries Inc.
  d Includes awards previously reported separately as BDM International Inc.
  f Listed previously as United Defense Limited Partnership and FMC Corp.
  g Includes awards previously reported as Allied Signal Inc.
  h Includes awards previously reported as General Electric Co. PLC.
  i Includes awards previously reported as Mobil Corp. and Exxon Corp.
  j Includes awards previously reported as MCI Communications Corp.

NATIONAL AERONAUTICS AND SPACE ADMINISTRATION
MAJOR CONTRACTORS

Fiscal Years 1998–2001
Listed by rank according to net value of prime
contracts awarded during last fiscal year
(Millions of Dollars)

Company	1998	1999	2000	2001
TOTAL PROCUREMENTS..............	$12,561	$12,675	$12,505	$12,748
Awards to Business Firms..............	9,551	9,386	9,273	9,210
% of TOTAL PROCUREMENTS	76%	74%	74%	72%
United Space Alliance LLC	$ 1,480	$ 1,465	$ 1,609	$ 1,659
The Boeing Co.	1,488	1,205	1,236	952
Lockheed Martin Corp.	982	906	710	608
Lockheed Martin Space Operations Co.	36	296	485	494
Thiokol Corp.	364	395	368	378
Boeing North America	261	272	258	304
McDonnell Douglas Corp.	420	416	320	282
Space Gateway Support	(a)	221	218	261
Lockheed Martin Engrg. & Science ...	227	231	287	228
Science Applications Int'l Corp.	78	91	107	138
Raytheon Information Systems Co. ...	92 [b]	123 [b]	130	128
Computer Sciences Corp.	177	174	143	126
QSS Group Inc.	13	22	83	126
Swales & Associates Inc.	38	57	63	116
Hughes Aircraft Co.	108	174	88	104
United Technologies Corp.	91	104	107	89
Raytheon Technical Services Co. ...	45 [c]	49	62	86
TRW Inc...................................	224	184	124	83
Ball Aerospace & Tech. Corp.	69	63	67	81
Orbital Sciences Corp.	57	73	82	74
Sverdrup Technology Inc...............	29	58	74	70
Delta Launch Services	(a)	(a)	(a)	62
Honeywell Technology Solutions ...	275	85	67	60 [d]
Mississippi Space Services	(a)	16	57	59
Science Systems Applications	15	19	20	57
Intellisource Information Systems ...	(a)	20	18	56
Hamilton Sundstrand Space Systems	55 [f]	58	55	55
Wyle Laboratories	42	46	55	53
ITT Corp....................................	57	47	45	48
Johnson Engineering Corp.	54	58	54	45

Source: National Aeronautics and Space Administration, "Annual Procurement Report" (Annually).
   a   Not in list of major contractors for indicated year(s).
   b   Previously reported as Hughes Information Technology Corp.
   c   Previously reported as Hughes Training Inc.
   d   Previously reported as Allied Signal Technical Services.
   f   Previously reported as Hamilton Standard Space Systems.

DEPARTMENT OF DEFENSE PRIME CONTRACT AWARDS OVER $25,000
FOR SELECTED MAJOR MILITARY HARD GOODS
BY GEOGRAPHIC REGION
Fiscal Years 1999, 2000, and 2001

Program and Region	Millions of Dollars			Percent of Program Total		
	1999	2000	2001	1999	2000	2001
AIRCRAFT—TOTAL	$25,673	$33,285	$33,491	100.0%	100.0%	100.0%
New England	$ 1,890	$ 2,094	$ 3,319	7.4%	6.3%	9.9%
Middle Atlantic	1,657	2,663	2,258	6.5	8.0	6.7
East North Central	1,467	2,936	3,163	5.7	8.8	9.4
West North Central	4,276	4,373	4,560	16.7	13.1	13.6
South Atlantic..............	5,818	5,525	7,423	22.7	16.6	22.2
East South Central	526	814	734	2.0	2.4	2.2
West South Central	3,212	7,001	4,257	12.5	21.0	12.7
Mountain....................	1,581	1,792	1,888	6.2	5.4	5.6
Pacific[a]	5,248	6,087	5,889	20.4	18.3	17.6
MISSILE & SPACE SYSTEMS—TOTAL.........	$11,162	$ 9,444	$ 9,347	100.0%	100.0%	100.0%
New England	$ 589	$ 611	$ 355	5.3%	6.5%	3.8%
Middle Atlantic	309	390	494	2.8	4.1	5.3
East North Central	175	146	61	1.6	1.5	0.7
West North Central	195	139	149	1.7	1.5	1.6
South Atlantic..............	1,237	1,149	596	11.1	12.2	6.4
East South Central	583	446	594	5.2	4.7	6.4
West South Central	1,219	512	416	10.9	5.4	4.4
Mountain....................	2,693	2,309	2,773	24.1	24.5	29.7
Pacific[a]	4,163	3,742	3,911	37.3	39.6	41.8
ELECTRONICS & COMMUNICATIONS EQUIPMENT—TOTAL ...	$12,949	$11,916	$13,875	100.0%	100.0%	100.0%
New England	$ 1,593	$ 1,329	$ 1,860	12.3%	11.2%	13.4%
Middle Atlantic	1,736	1,677	1,808	13.4	14.2	13.0
East North Central	666	696	887	5.1	5.8	6.4
West North Central	573	710	741	4.4	6.0	5.3
South Atlantic..............	4,691	4,104	4,433	36.2	34.4	31.9
East South Central	226	202	204	1.7	1.7	1.5
West South Central	872	819	857	6.7	6.9	6.2
Mountain....................	621	622	845	4.8	5.2	6.1
Pacific[a]	1,971	1,758	2,240	15.2	14.8	16.1

Source: Department of Defense, Washington Headquarters Services, Directorate for Information Operations and Reports.
a   Includes Alaska and Hawaii.

GLOSSARY

Aeronautics: the science and art of designing and constructing aircraft, also, the art or science of operating aircraft.

Aerospace Employment: annual average calculated as one-twelfth of sum of monthly estimates of total number of persons employed during a designated pay period by the aircraft, missile, and space industries (SICs 372 and 376) plus estimated aerospace-related employment in the communications equipment (SIC 3662), instruments (SICs 381 and 382), and in certain other industries (SICs 28, 35, 73, 89, etc.)

Aerospace Industry: the industry engaged in research, development, and manufacture of aerospace systems including: manned and unmanned aircraft; missiles; spacecraft; space launch vehicles; propulsion, guidance, and control units for all of the foregoing; and a variety of airborne and ground-based equipment essential to the test, operation, and maintenance of flight vehicles.

Aerospace Payroll: estimated on the basis of average weekly *earnings* for a given calendar year for *production workers* plus an estimated annual salary for other employees.

Aerospace Sales: the *AIA* estimate of *aerospace industry sales,* developed by summing: *DoD* expenditures for *aircraft, missiles,* and space-related *procurement* and *RDT&E; NASA* expenditures for *research and development* and space flight control and data communications; *outlays* for space activities by other U.S. government departments and agencies; commercial sales of space-related products; net domestic and export sales of civil aircraft, engines, and parts; *Foreign Military Sales* and commercial exports of military aircraft, missiles, propulsion, and related parts; sales of *related products and services* including: electronics, software, and ground support equipment; and sales of *non-aerospace products* which are produced in aerospace-manufacturing *establishments* and which use technology, processes, and materials derived from the aerospace industry.

AIA: Aerospace Industries Association of America, Inc., formerly Aircraft Industries Association.

Air Carriers: the commercial system of air transportation, consisting of domestic and international scheduled and charter service.

Aircraft: all airborne vehicles supported either by buoyancy or by dynamic action. Used in this volume in a restricted sense to mean an airplane—any winged aircraft including helicopters, but excluding gliders and guided missiles.

Aircraft Agreement (Agreement on Trade in Civil Aircraft): negotiated in the Tokyo Round of the *Multilateral Trade Negotiations* and implemented January 1, 1980, providing for elimination of tariff and non-tariff trade barriers in the civil aircraft sector.

Aircraft Industry: the industry primarily engaged in the manufacture of aircraft, aircraft engines, and parts including propellers and auxiliary equipment. A sector of the *Aerospace Industry.*

Airframe: the structural components of an airplane, such as: fuselage, empennage, wings, landing gear, and engine mounts, but excluding such items as: engines, accessories, electronics, and other parts that may be replaced from time to time.

Airlines: see *Air Carriers.*

Appropriation (Federal Budget): an act of Congress authorizing an agency to incur *obligations* and make payments out of funds held by the Department of the Treasury.

Assets, Net: the sum of all recorded assets after reducing such amount by allowance of reserve for bad debts, *depreciation,* and amortization, but before deducting any liabilities, mortgages, or other indebtedness.

Astronautics: the art and science of designing, building, and operating manned or unmanned space objects.

Average Weekly Hours: average hours for which pay was received; different from standard or scheduled hours.

Avionics: communications, navigation, flight controls, and displays.

Backlog: the *sales* value of *orders* accepted (supported by legal documents) that have not yet passed through the sales account.

BMDO: Ballistic Missile Defense Organization, an agency of the Department of Defense.

Budget Authority: authority provided by the Congress; mainly in the form of *Appropriations,* which allows Federal agencies to incur *obligations* to spend or lend money.

Bureau of Economic Analysis (BEA): an agency of the Department of Commerce.

Bureau of Labor Statistics (BLS): an agency of the Department of Labor.

Bureau of the Census: an agency of the Department of Commerce.

Constant Dollars: calculated by dividing current ("then-year") dollars by appropriate price *deflator* and multiplying the result by 100.

Deflator: index used to convert a price level to one comparable with the price level at a different time, offsetting the effect of inflation. The base period, which equals 100, is usually specified as either a given fiscal or calendar year.

Depreciation: the general conversion of the depreciable cost of a fixed asset into expense, spread over its remaining life. There are a number of methods, all based on a periodic charge to an expense account and a corresponding credit to a reserve account.

Development: the process or activity of working out a basic design, idea, or piece of equipment. See also *Research and Development.*

DoD: Department of Defense.

DoE: Department of Energy.

DoT: Department of Transportation.

Durable Goods Industry: comprised of major manufacturing industry groups with SIC Codes 24, 25, and 32-39. All major manufacturing industry groups in SIC Codes 20-23 and 26-31 are considered nondurable goods manufacturing industry groups.

Earnings: the actual return to the worker for a stated period of time. Irregular bonuses, retroactive items, payments of various welfare benefits, and payroll taxes paid by employers are excluded.

Average Hourly Earnings: on a "gross" basis, reflecting not only changes in basic hourly and incentive wage rates, but also such variable factors as: premium pay for overtime, late shift work, and changes in output of workers paid for an incentive plan.

Average Weekly Earnings: derived by multiplying *average weekly hours* by *average hourly earnings.*

Establishment: the basis for reporting to the Census of Manufacturers; an operating facility in a single location.

Evaluation (Department of Defense): determination of technical suitability of material, equipment, or a system. See *RDT&E*.

Expenditures (Federal Budget): see *Outlays*.

Export–Import Bank of the United States (Eximbank): created in 1934 and established as an independent U.S. government agency in 1945, Eximbank is designed "... to aid in financing and to facilitate *exports...*" Eximbank receives no *appropriations* from the U.S. Congress. It is directed by statute to: (1) offer financing that is competitive with that offered exporters of other countries by their official export credit institutions, (2) determine that the transactions supported provide for a reasonable assurance of repayment, (3) supplement, but not compete with private sources of export financing, and (4) take into account the effect of its activities on small business, the domestic economy, and U.S. employment.

Exports: domestic merchandise including commodities which are grown, produced, or manufactured in the United States and commodities of foreign origin which have been changed in the United States from the form in which they were imported or which have been enhanced in value by further manufacture in the United States and which are traded or sold to other nations.

FAA: Federal Aviation Administration (formerly the Federal Aviation Agency), an agency of the Department of Transportation.

Facility: a physical plant or installation including: real property, building, structures, improvements, and plant equipment.

Fiscal Year (Federal Budget): beginning October 1, 1976, the fiscal years run from October 1 through September 30 and are designated by the year in which they end.

Flyaway Value: includes the cost of the *airframe,* engines, electronics, communications, armament, and other installed equipment.

Footnotes: common to many tables throughout this edition are the following:

E	Estimate.
NA	Not available/ Not applicable.
p	Preliminary.
r	Revised.
Tr.Qtr.	*Transition Quarter.*

NOTE: Detail may not add to totals because of rounding.

Foreign Military Sales (FMS): export *sales* to foreign governments arranged through the Department of Defense, whereby *DoD* recovers full purchase price and administrative costs; often mistakenly used to include foreign military aid and foreign commercial sales as well.

FY: see *Fiscal Year.*

GDP (Gross Domestic Product): the market value of goods and services produced by labor and property located in the United States.

General Agreement on Tariffs and Trade (GATT): a multilateral treaty among more than 100 governments whose primary mission is the reduction of trade barriers. The World Trade Organization was established January 1, 1995 to implement the agreement and provide a forum to discuss trade issues.

General Aviation: all civil flying except that of *air carriers.*

Helicopter: a rotary-wing *aircraft* which depends principally for its support and motion in the air upon the lift generated by one or more power-driven rotors, rotating on substantially vertical axes. A helicopter is a *V/STOL.*

Heliport: an area, either at ground level or elevated on a structure, that is used for the landing and take-off of helicopters and includes some or all of the various facilities useful to *helicopter* operations such as: helicopter parking, hangar, waiting room, fueling, and maintenance equipment.

Helistop: a minimum facility *heliport,* either at ground level or elevated on a structure for the landing and takeoff of helicopters, but without such auxiliary facilities as: waiting room, hangar parking, etc.

ICBM: InterContinental Ballistic Missile, with a range of more than 5,000 miles.

Imports: classified as "general imports" or "imports for consumption." This volume refers generally to "imports for consumption," which are entries for immediate consumption plus merchandise withdrawn from bonded storage warehouses for consumption. Data are compiled from Import Entries filed with U.S. Customs officials and are in general based on the market value or price in the foreign country at the time of exportation of such merchandise, including the cost of containers and coverings, as well as other charges and expenses incidental to placing the merchandise in condition, packed and ready for shipment to the United States, but excluding import duties, insurance, freight, and other charges incidental to arrival of the goods in the United States. The foreign values of imported merchandise are converted into U.S. currency at the rate of exchange prevailing on the day the merchandise is shipped to the United States.

Income:

Net Operating Income: total *sales* less total operating costs.

Other Income and Expenses: includes interest income, royalty income, capital gains and losses, interest expense, cash discounts, etc.

Net Income (Before Income Taxes): *Net Operating Income* plus or minus *Other Income and Expenses.*

Net Income (After Income Taxes): *Net Income (Before Income Taxes)* less federal income taxes.

Lump-Sum Wage Payment: a one-time payment given in lieu of general wage increases and/or cost of living adjustments in labor settlements.

Manufacturing Industries: those *establishments* engaged in the mechanical or chemical transformation of inorganic or organic substances into new products, and usually described as plants, factories, or mills, which characteristically use power-driven machines and materials-handling equipment; also establishments engaged in assembling component parts of manufactured products if the new product is neither a structure nor other fixed improvement.

Merchandise Trade Balance: the difference between the value of U.S. goods exported to other countries and foreign goods imported into this country. The trade balance is generally regarded as "favorable" when *exports* exceed *imports*—a trade surplus—and "unfavorable" when imports exceed exports—a trade deficit.

Missile: sometimes applied to space launch vehicles, but more properly connotes automated weapons of warfare, that is, a weapon which has an integral system of guidance, as opposed to the unguided rocket.

Multilateral Trade Negotiations (MTN): a forum within the *GATT* in which countries negotiate to overcome their trade problems. Awaiting ratification by each of the 112 nations involved in the MTN, the "Uruguay Round" seeks to strengthen the GATT and expand its disciplines to new areas such as: services, agriculture, and trade-related intellectual property rights.

NAICS (North American Industrial Classification System): a system developed by Canada, Mexico, and the U.S. government that groups *establishments* into industries based on a production-oriented concept in order to provide uniformity and comparability of statistical data and facilitate economic analyses between industries and the three North American countries.

NASA: National Aeronautics and Space Administration.

NATO: North Atlantic Treaty Organization.

New Obligational Authority (Federal Budget): see *Budget Authority*.

Non-Aerospace Products and Services: products and services other than *aircraft, missiles, space vehicles,* and related propulsion and parts, produced or performed by *establishments* whose principal business is the development and/or manufacture of aerospace products.

OASD: Office of the Assistant Secretary of Defense.

Obligations (Federal Budget): commitments made by Federal agencies to pay out money for products, services, or other purposes—as distinct from the actual payments. Obligations incurred may not be larger than *budget authority*.

Orders, Net New: the *sales* value of new orders (supported by legal documents) minus cancellations during the period.

Other Aerospace Products and Services: all conversions, modifications, site activation, other aerospace products (including drones), services, plus *research and development* under contract, defined as: *basic* and *applied research* in the sciences and in engineering and design and *development* of prototype products and processes.

Other Customers: all customers other than the U.S. government to include but not limited to: *air carriers,* private citizens and corporations, and state, local, and foreign governments.

Outlays: checks issued, interest accrued on the public debt, or other payments made, net of refunds and reimbursements.

Overtime Hours: that portion of the gross *average weekly hours* which was in excess of regular hours and for which premium payments were made.

Passenger-Mile: one passenger moved one mile.

Payroll, All Manufacturing: includes the gross *earnings* paid in the calendar year to all employees on the payroll of operating manufacturing *establishments*. Includes all forms of compensation paid directly to workers such as: salaries, wages, commissions, dismissal pay, all bonuses, vacation and sick leave pay, and compensation in kind; prior to such deductions as: employees' Social Security contributions, withholding taxes, group insurance, union dues, and savings bonds. Does not include employers' Social Security contributions or other non-payroll labor costs such as: employees' pension plans, group insurance premiums, and workmen's compensation.

Procurement: the process whereby the executive agencies of the Federal Government acquire goods and services from enterprises other than the Federal Government.

Production Workers: includes working foremen and all non-supervisory workers (including lead-men and trainees) engaged in fabricating, processing, assembling, inspection, receiving, storage, handling, janitorial services, product development, auxiliary production for plant's own use, and record-keeping and services closely associated with the above production operations.

RDT&E (Department of Defense): Research, Development, Test, and Evaluation.

Related Products and Services: sales of electronics, software, and ground equipment in support of aerospace products, plus sales by aerospace manufacturing *establishments* of systems and equipment which are generally derived from the industry's aerospace technological expertise in design, materials, and processes, but which are intended for applications other than flight.

Research: see *Research and Development.*

Research and Development:

Research: systematic study directed toward fuller scientific knowledge or understanding of the subject studied. Research is classified as either basic or applied according to the objectives of the sponsoring agency.

Applied Research: with the objective of gaining knowledge or understanding necessary for determining the means by which a recognized and specific need may be met.

Basic Research: with the objective of gaining fuller knowledge or understanding of the fundamental aspects of

phenomena and of observable facts without specific applications toward processes or products in mind.

Development: the systematic use of scientific knowledge directed toward the production of useful materials, devices, systems, or methods including design and development of prototypes and processes.

Independent Research and Development (IR&D): a term devised by the Department of Defense and used by Federal agencies to differentiate between a contractor's research and development technical effort performed under a contract, grant, or other arrangement (R&D) and that which is self-initiated and self-funded (IR&D).

Industrial Research and Development: research and development work performed within company facilities, funded by company or Federal funds, and excluding company-financed research and development contracted to outside organizations such as: research institutions, universities and colleges, or other non-profit organizations.

Rotorcraft: an *aircraft* which, in all its usual flight attitudes, is supported in the air wholly or in part by a rotor or rotors (i.e., airfoils rotating or revolving about an axis). See *Helicopter.*

Sales: net of returns, allowances, and discounts, the dollar value of shipments, including dealer's commissions, if any, which have passed through the sales account.

Satellite: a body that revolves around a larger body, such as the Moon revolving around the Earth, or a man-made object revolving about any body such as the Sun, Earth, or Moon.

SIC (Standard Industrial Classification): a system developed by the U.S. government to define the industrial composition of the economy, facilitating comparability of statistics. See *Aerospace Industry* for explanation of SIC codes applicable to the aerospace industry.

Space Vehicle: an artificial body operating in outer space (beyond the Earth's atmosphere).

Stockholder's Equity: *assets* minus all obligations of the corporation, except those to stockholders. Annual data are average equity for the year (using four end-of-quarter figures). For details, see "Quarterly Financial Report for Manufacturing, Mining and Trade Corporations," compiled by the *Bureau of the Census.*

STOL: short take-off and landing *aircraft.*

Test (Department of Defense): an experiment designed to assess progress in attainment or accomplishment of *development* objectives (see *RDT&E*).

Thrust: the driving force exerted by an engine, particularly an *aircraft* or *missile* engine, in propelling the vehicle to which it is attached.

Ton-Mile: one ton moved one mile.

Total Obligational Authority: the sum of *budget authority* granted or requested from the Congress in a given year, plus unused budget authority from prior years.

Trade Balance: see *Merchandise Trade Balance.*

Transition Quarter (Tr. Qtr.): the three-month interval from July 1, 1976 to September 30, 1976 belonging to neither Fiscal Year 1976 nor Fiscal Year 1977. See *Fiscal Year.*

Turbine, Turbo: a mechanical device or engine that spins in reaction to a fluid flow that passes through or over it. Frequently used in "turboprop" or "turbo-jet."

UK: United Kingdom.

US: United States of America.

USA: United States Army, an agency of the U.S. Department of Defense.

USAF: United States Air Force, an agency of the U.S. Department of Defense.

USN: United States Navy, an agency of the U.S. Department of Defense.

USSR: Union of Soviet Socialist Republics. Statistics continue to exclude this region until official data from the now independent republics become available.

Utility Aircraft: an aircraft designed for general purpose flying.

V/STOL: vertical short take-off and/or landing *aircraft.*

INDEX

INDEX

171